Black Faced Sheep

Their History, Distribution and Improvement with Methods of Management

by John and Charles Scott

with an introduction by Jackson Chambers

This work contains material that was originally published in 1888.

Self Reliance Books

Get more historic titles on animal and stock breeding, gardening and old fashioned skills by visiting us at:

http://selfreliancebooks.blogspot.com/

Introduction

I am pleased to present yet another practical title on breeding and raising livestock.

The work is in the Public Domain and is re-printed here in accordance with Federal Laws.

As with all reprinted books of this age that are intended to perfectly reproduce the original edition, considerable pains and effort had to be undertaken to correct fading and sometimes outright damage to existing proofs of this title. At times, this task is quite monumental, requiring an almost total "rebuilding" of some pages from digital proofs of multiple copies. Despite this, imperfections still sometimes exist in the final proof and may detract from the visual appearance of the text.

I hope you enjoy reading this book as much as I enjoyed making it available to readers again.

Jackson Chambers

Yours faithfully
C. Howatson

TO

CHARLES HOWATSON, Esq.,

OF GLENBUCK;

BY

WHOSE ENTHUSIASTIC AND UNTIRING EFFORTS,

AS A BREEDER AND EXHIBITOR,

THE BLACKFACED SHEEP OF SCOTLAND

HAVE BEEN SIGNALLY IMPROVED AND RAISED IN POPULARITY;

𝕿𝖍𝖎𝖘 𝖁𝖔𝖑𝖚𝖒𝖊

IS RESPECTFULLY INSCRIBED

BY

THE AUTHORS.

PREFACE.

———◆———

In the following pages we have dealt as much as possible with useful facts. The origin of the Blackfaced Sheep of Scotland, which is lost in obscurity, is of less interest and importance than the knowledge that the breed has retained its purity from time immemorial, and that its improvement since the end of last century has been a gradual process, brought about solely by careful breeding and selection, without crossing or intermixture of races. In weight and quality of both mutton and wool, the value of the Black-faced sheep has almost doubled within the memory of man; and in point of early maturity the improvement is greater, for one-year-old wether mutton is now taking the place of the four- or five-years-old mutton, which was the rule not so many years ago.

There is little doubt that a great deal of the improvement noted is due to the friendly rivalry of breeders in the show-yard. This is all the more noticeable because there has been less over-feeding for show with this breed than any other; while it is the only breed shown in the natural fleece, without any colouring of the wool, and without any attempts at clipping into shape. The result has been to immensely popularise the show system with Blackfaced breeders.

Within the last few years, indeed, the leading men amongst them have not only made the Blackfaces the most prominent exhibit at nearly all the principal Scotch shows, but have placed the breed in quite a unique position. It is the only breed in Scotland for which there are classes for ram lambs, and the only breed for which prizes have as yet been offered for single animals in the ewe and ewe-hogg classes. The spirit of improvement evinced in these new movements cannot fail to raise the Blackfaced sheep to a yet higher pinnacle of fame.

In addition to the characteristics of the breed, we have given a scale of points, with definite values attached, which it is hoped may not only be a help to young breeders in their efforts to improve their flocks, but may yet form the basis of a standard for judging Blackfaces. It is of the first importance that correct judgment should be given at shows. The whole influence is wielded by the judges, and if they place it in the wrong scale the whole labour of the year is worse than lost, it becomes mischievous, and had better not have been.

The annual ram sales afford another striking illustration of the progress of this breed. It is not only that a far greater number of breeders are engaged in the trade, nor that a far greater number of Blackfaced rams are now used for breeding purposes than was the case a quarter of a century ago; the generally improved character of the lots offered for sale is equally remarkable. But most noteworthy of all is the fact that, in the face of falling markets, there has latterly been a great and steady rise in the general average price of rams sold. Thus, while in 1885 the 4381 rams sold only averaged £2, 15s. 4d., this year 4952 made an average price of £4, 5s. 10d., being a rise of £1, 10s. 6d., nearly 55 per cent., in two years of severe agricultural depression.

The Blackfaces have never been so numerous nor so widely distributed as at the present time. We meet with the breed all over Scotland and the north of England, in parts of Wales and Ireland, and even in America, New Zealand, and other countries abroad. In Scotland alone the Blackfaces number at least four and a quarter millions, or fully 70 per cent. of the total sheep stock of the country, representing an invested capital of about £7,634,065, and an annual rental of £1,177,215.

If this great industry is to flourish, the efforts of breeders must be seconded, and existing hindrances in the shape of over-costly entries and unequable valuations remedied. At the same time, the already improved character of the Blackfaced sheep must be upheld, and, if possible, made better. A larger percentage of lambs must also be reared. All this demands more attention to the details of management, including better wintering, the subdivision of the pastures, and, where the grazings are wet and exposed, improved drainage and shelter. These, in our opinion, are the cardinal points on which hinges the future development and success of Blackfaced sheep-farming.

CONTENTS.

———◆◆———

CONTENTS.

ILLUSTRATIONS.

The Celebrated Blackfaced Ram "CAIRNTABLE,"
Bred by CHARLES HOWATSON, of Dornel, and sold for **SIXTY POUNDS**,
To J. N. FLEMING, of Keil.   July, 1870.

Blackfaced Ram "GLENBUCK AGAIN." Prize Lamb at H. & A. S. Show, Glasgow, 1882; 2nd prize Shearling at H. & A. S. Show, Inverness, 1883; aged Ram in 1st prize male family group at H. & A. S. Centenary Show, 1884; sire of 1st prize Shearling Ram at H. & A. S. Show, Aberdeen, 1885.

"Highland Mary."   "Centenary Champion."   "Seventy-Two."

BLACKFACED RAM "GLENBUCK YET."

To face page 39.]

To face page 16.]

BLACKFACED RAM "SEVENTY-TWO."

[See page 22.

BLACKFACED SHEEP.

CHAPTER I.

HISTORY OF BLACKFACED SHEEP.

THE origin of the Scotch blackfaced sheep is shrouded in mystery. Whence the breed originally sprang has not been revealed in any of the historical records yet brought to light. Several theories, however, have been advanced as to the possible origin of the breed. One is, that the blackfaced sheep are the native original stock of the country. Another maintains that they were introduced into Ettrick Forest by James IV., King of Scotland, in 1503; but the old chronicles omit to tell us where the royal husbandman secured the flock of 20,000 sheep of this breed which he is said to have brought into Selkirkshire, though they are supposed, with small reason perhaps, to have been brought from Fifeshire. A race of sheep similar to the Scotch blackfaced is said to have existed in Yorkshire at a very early period, and it is, of course, possible that the King may have brought them from that quarter; but here again no evidence exists to prove that he did so. Again, tradition asserts that the breed was originally confined to various districts in Scotland, where it was known by the several names of "Linton," "Forest," "Tweeddale," and "Lammermuir."

Naismyth of Hamilton, writing in 1796 of a visit he made

A

to Lammermuir, states that the breed prevalent there was the blackfaced muir kind, having generally horns, and called the "short" sheep, but that "it is impossible to trace their origin, there being no tradition of the sheep here ever being of a different kind; nor can they be called a distinct variety of the species, for a considerable difference of figure and fleece may be discovered among the individuals, even of the flocks to which the greatest attention has been paid."

Louden's "Cyclopædia" states that the *dunfaced* breed—which is only another name for blackfaced—"said to have been imported into Scotland from Denmark or Norway at a very early period, still exists in most of the counties to the north of the Frith of Forth, though only in very small flocks. Of this ancient breed there are now several varieties, produced by peculiarities of situation and different modes of management, and by occasional intermixture with other breeds."

In a report on the agriculture of Peebles, written by the Rev. Mr. Findlater in 1802, it is stated that "no clear tradition nor even conjecture can be given as to when or whence sheep were first introduced into this county, or whether the present breed are indigenous or from another country. There is, indeed, an obscure tradition, that previous to the introduction or general prevalence of sheep in the parish of Tweedsmuir, the farmers in that parish paid their rents by grazing, for hire through the summer, the oxen then generally used by Lothian farmers for their winter ploughing. The native Tweeddale breed, which has continued the same as far back as memory or tradition extends, are all horned, with black faces and black legs and coarse wool."

Galloway has also been credited as the original home of the breed. In the old statistical account of Scotland the Rev. James Muirhead, who wrote the report on the parish of Urr, fixed the date of the introduction of blackfaced sheep as about the year 1603. But he asks whence these sheep came. "It may be observed that Galloway abounds with goats, which in the marshy or soft tracts are almost entirely of a black colour." This writer then gives some countenance to the theory that the goats and sheep bred together, mentioning that crosses between the two animals were quite common. But while venturing on this suggestion, Muirhead confesses that any inquiry on this subject is not attended with much satisfaction.

At the present day Mr. John Fleming, Ploughland, keeps several he-goats to go amongst the sheep to tup ewe hoggs—and he finds that goats and sheep will not breed.

Further examination of the recorded opinions of many writers leads to no more satisfactory result. So much is given as conjecture, that it is quite impossible to form any definite conclusion of the origin of the breed, based on reliable authenticity. All that can be ascertained is, that from time immemorial blackfaced sheep have been settled in the mountainous districts of the south of Scotland, and they are supposed to be the direct descendants of the aboriginal Scotch sheep, their present improvement having been brought about by a long-continued and judicious process of selection.

If it be true—and there is no reason to think otherwise—that no foreign blood has been introduced into the breed since the time when the old Scotch variety called "short" or "dunfaced" sheep existed, it is certain that the blackfaced breed, as it now exists, is the oldest variety known in Great Britain. Crossing with other races, although it has been attempted, has never succeeded ; and although the breed is now very different in point of quality to what it was formerly, it can lay claim to an unbroken line of pedigree many centuries old. Hector Bœthius, who wrote about the year 1460, speaking of the sheep in the vale of Esk, says :—

"Until the introduction of the Cheviots, the rough-woolled blackfaced sheep alone were to be found." From this it would appear that the blackfaces had been the prevailing breed at an early period; and if this statement can be credited, the previous one about King James introducing the breed into Scotland about the year 1503 is plainly contradicted. As already mentioned, the origin of the breed is shrouded in mystery, and the traditionary evidence here and elsewhere related can only be accepted as untrustworthy in point of fact.

The introduction of blackfaces into the Highlands of Scotland is of a more recent date, and can be traced with greater accuracy. Their first appearance is mentioned in reports on the counties of Dumbarton and Perthshire as about the year 1750. From Perthshire they were taken to the northern counties by Sir John Lockhart of Balnagown. They were first introduced into the West Highlands about the year 1762 by a Mr. Campbell, at one time proprietor of Garieve, Ayrshire, and were soon found far more profitable than cattle for the higher parts of the country, to which they afterwards rapidly spread throughout the whole of Scotland. It does not belong to this part of the work to give any account of the change which this stock produced upon the population of the Highlands, and space does not permit us to trace its progress from one county to another. "One thing is indisputable," says Sir John Sinclair; "a much greater value of mutton and wool than of beef was speedily obtained from the higher grounds, and all those situations where little or no provision could be made for cattle in winter. The best interests of the nation at large, as well as of proprietors, were therefore promoted by the change."

Shortly after the beginning of the present century a popularity sprang up in favour of the Cheviot breed. The difference in the value of wool told greatly in favour of the "white," and to such an extent were the blackfaces supplanted, that the breed came very nearly being totally extinguished. In reference to this point, a writer, in describing a tour from Land's End to John o'Groat's, speaking of the upper reaches of Strathspey, says:—"The sheep in this region are chiefly the old Scotch breed, with curling horns and crooked faces and legs, such as are represented in old pictures. The black seems to be spattered upon them, and looks as if the heather would rub it off. The wool is long and coarse, giving them a goat-like appearance. They seem to predominate over any other breed in this part of the country, yet not necessarily nor advantageously. A large sheep-farmer from England was staying at the inn, with whom I had much conversation on the subject. He said the Cheviots were equally adapted to the Highlands, and thought they would ultimately supplant the blackfaces. Although he lived in Northumberland, full two hundred miles to the south, he had rented a large sheep-walk or mountain farm in the Western Highlands, and had come to this district to buy or hire another tract. He kept about 4000 sheep, and intended to introduce the Cheviots upon the Scotch holdings, as their bodies were much heavier and their wool worth nearly double that of the old blackfaced breed. Sheep are the principal source of wealth in the whole of the north and west of Scotland. I was told that sometimes a flock of 20,000 is owned by one man. The lands on which they are pastured will not rent above one or two shillings *per acre;* and a flock of even 1000 requires a vast range, as may be indicated by the reply of a Scotch farmer to an English one, on being asked by the latter, 'How many sheep do you allow to the acre?' 'Ah, mon,' was the answer, 'that's nae the way we counts in the Highlands; it's how monie acres to the sheep.'"

A few severe winters, however, soon revealed the fact that the Cheviot breed was unable to exist in exposed situations, and the blackfaces were again reinstated to at least all the higher grazings in the country. For many years after this, perhaps quite half a century, little changing took place in the positions of the two breeds. Farms as well as farmers were known to be inseparably associated with one or other of the breeds, and in this way many years intervened. For the while, however, the Cheviots reigned supreme. But about the year 1860, which happened to be a very severe one on hill flocks, the Cheviots began to decline in favour among hill farmers. Every year following saw many old flocks of that breed displaced by the hardier blackfaces, and the changes which have taken place within the last few years have been not a little extraordinary. This may be attributed partially to the Cheviots having been bred too tender in constitution, but chiefly owing to the great improvement recently effected in the quality of the blackfaces.

Coming to the more immediate history of the blackfaces, the fame which they have now acquired is due to breeders of the present generation. M'Kersie of Glenbuck, Dun of Kirton, in the parish of Campsie, the Weirs of Priesthill, and Gillespies of Douglas Water, Lanarkshire, were perhaps the earliest improvers; but the breed owes most of its present popularity to such men as Messrs. Howatson of Glenbuck; Archibald, Overshiels; Foyer, late of Knowhead; Welsh, Earlshaugh; and others.

The following list gives, as nearly as possible, all the breeders who have exhibited blackfaced sheep at shows or ram sales during the past year :—

Allan, G., Whiteleehill, Ayr.
Anderson, James, Greenockdyke, Muirkirk.
Anderson, William, Whithaugh, Muirkirk.
Archibald, J., Overshiels, Stow, Midlothian.
Argyll, Duke of, Ballymenach, Campbeltown.
Athole, Dowager-Duchess of, St. Colmes.
Aveland, Lord, Corrychrone.

Blackwood, William, Carskeoch, Patna.
Boyd, Col. Hay, Dornel.
Breadalbane, Marquis of, Taymouth.
Brown, J., Lochbrowan, Ayr.
Bryce-Buchanan, ——, Bolquhan.
Bryden, ——, Burn Castle.
Buchanan, Robert, Letter, Killearn.
Buchanan, D. M. L. B., Bolquhan, Killearn.
Bute, Marquis of.
Byres, ——, Baadsmill.
Cadzow, C., Borland, Biggar.
Cadzow, R. & T., Borland, Biggar.
Callan, William, Crossflat, Muirkirk.
Campbell, Donald, Sheraston, Crieff.
Cecil, Lords A. & L., Orchardmains, Innerleithen.
Clark, Alexander, Todlaw, Lesmahagow.
Clark, A., Todlaw.
Clark, Wm., Auchinlongford.
Clarkson, J. & A., Prett's Mill, Lanark.
Cowan, Alex., Spittalhill, Fintry.
Cowan, Andrew, Spittalhill, Fintry.
Cowan, Mrs., Lurg, Fintry.
Cowbrough, John, Blairtumnock, Campsie.
Craig, A., Bankend, Strathaven.
Craig, Daniel, Middlefield, Muirkirk.
Craig, D., Priesthill.
Craig, D., Glenglass.
Craig, John, High Ploughland, Strathaven.
Craig, John, Invergeldie, Comrie, Crieff.
Craig, John, Southhalls, Strathaven.
Craig, J., Representatives of the late, Craigdarroch, Sanquhar.
Craig, Robert, Netherwood, Muirkirk.
Cunningham, Alex., Clews, Douglas.
Dempster, Captain, Ladyton, Galston.
Donald, J., White Clauchrie, Ayr.
Donald, W., Hairshaw, Strathaven.
Duncan, J., of Benmore, Greenock.
Ewing, Sir A. Orr, M.P., Ballikinrain Castle.
Fleming, H., Lochfield, Newmilns.
Fleming, H., Overhouses, Strathaven.
Fleming, John, Meadowbank, Strathaven.
Fleming, John, Ploughland, Strathaven.

Fleming, ——, Westbrown Castle.
Fleming, ——, Threipland.
Fletcher, Angus, Auchtertyre, Tyndrum.
Forest, C. Low, Ploughland, Strathaven.
Gemmell, G. A., Garpel, Ayr.
Gilbert, Thomas, Walston, Biggar.
Gillespie, Jas. J., St. Colmes, Ballinluig, Perthshire.
Gordon, James A., of Arabella, Ross-shire.
Gordon, ——, Glenmeanie.
Gray, W. W., of Munraw, Haddington.
Gray, ——, Harperrig.
Greenshields, D., Garvald, Dolphinton.
Greenshields, James, West Town, Lesmahagow.
Hamilton, A., Drumclog, Strathaven.
Hamilton, G., Succoth, Arrochar.
Hamilton, Gavin, of Auldtown, Lesmahagow.
Hamilton, J., Woolfords, Carnwath.
Hamilton, John, Conenish, Tyndrum.
Hamilton, ——, Yardhouses.
Hazel, ——, Blackcraig, Ayr.
Hewetson, James, of Craiglearn, Auchenbainzie, Thornhill.
Hope, James, Midhouse, Muirkirk.
Hope, Thomas, Buroncastle, Strathaven.
Hope, William, High Ploughland, Strathaven.
Howatson, Charles, of Glenbuck, Glenbuck.
Howatson, W. M. S., Carskeoch, Patna.
Inglis, R., Loveston, Ayr.
Jack, R., Dykehead, Strathaven.
Jackson, Major Randle, Swordale, Evanton, Ross-shire.
Johnston, James, West Hope, Gifford.
Lamble, J., Gilbank, Strathaven.
Lawson, James, Lightshaw, Muirkirk.
Lees, R., Lagg, Ayr.
Leitch, ——, Glenboig.
Lind, Wm., South Cobbinshaw.
Lindsay, Wm., Craigend, Ayr.
Loudon, M., Meadowhead, East Kilbride.
Louden, J., Low Overmoor, Ayr.
Lovat, Lord.
Lumsden, James, of Arden House, Alexandria, N.B.
Malcolm, John, of Poltalloch.
Martin, D. T., of Girgenti, by Irvine.

Meikle, A., Juanhill, Strathaven.
Melrose, P., West Loch, Eddlestone.
Menzies, Sir Robt., Bart.
Menzies, Wm., Kielator, Tyndrum.
Milligan, ——, Kirkhope.
M'Cormick, ——, Lochenkit.
M'Culloch, J., Laggan, Ayr.
M'Diarmid, Duncan, Camusericht, Rannoch.
M'Dougall, Archibald, Claggan, Kenmore.
M'Gibbon, David, Ardnacraig, Campbeltown.
M'Gilchrist, ——, Fintry.
M'Indoe, Robt., Knowhead, Campsie.
M'Intyre, Donald, Tighnablair, Comrie.
M'Kinlay, Mrs., Maidencoates, Lanark.
M'Kean, ——, Bellewan.
M'Kersie, James, Cunningham House, Muirkirk.
M'Millan, J., Bent, Strathaven.
M'Min, ——, Meiklehill, Ayr.
M'Minn, T., Wellwood, Muirkirk.
M'Naughten, Alexander, Remony, Kenmore.
M'Pherson, R., Hapton, Ayr.
M'Turk, W. A., Barlae, Ayr.
Mitchell, William, Hazelside, Douglas.
Moffat, J. & J., Gateside, Sanquhar.
Mundell, Walter, Moy, Muir of Ord.
Mungle, ——, North Cobbinshaw.
Murray, John, Parkhall, Douglas.
Murray, James, Ploughland, Strathaven.
Murray, J., Midcrosswood, West Calder.
Murray, W. J., Mailinsland, Peebles.
Murray, ——, Eastside.
Norry, ——, Todholes.
Pate, James, Allerstocks, Strathaven.
Pate, D. B., Darnhaunch, Glenbuck.
Paterson, James, of Carmacoup, Douglas.
Paterson, John, Kirkton, Tyndrum.
Paterson, ——, Glentaggart, Representatives of the late.
Paul, Walter, Ibert, Killearn.
Paul, ——, Laighpark.
Pettigrew, Robt., Auldhouseburn, Muirkirk.
Pollock, J. J., of Auchineden, Strathblane.
Pringle, John, Castlemains, Douglas.

Purdie, John, Stonehill, Thankerton.
Rawlinson, R. & J., Docker Hall, Kendal.
Reid, James, Corsebank, Sanquhar.
Richmond, George, Drumshang, Maybole.
Ritchie, E., Stonehill, Crawfordjohn.
Robertson, Peter, Auchilty, Dingwall.
Rogerson, Thomas, Knowehead, Muirkirk.
Rollo, James, Boadhead, Dunning.
Roxburgh, J., Dimmurchie, Ayr.
Russel, James, Dundas Castle, South Queensferry.
Sample, J., Hairshawhead, Strathaven.
Sanderson, ——, Southtown.
Sandilands, J., Greens.
Sandilands, R., South Cumberland.
Scott, D., Laggish, Ayr.
Shaw, Thomas, Dunsyre, Biggar.
Sloan, John, Barnhill, Patna.
Stirling, James, Braco.
Shearer, G., North-Halls, Strathaven.
Smith, J., Mountainblow.
Stair, Earl of, Oxenfoord Castle, Dalkeith.
Steel, Captain, Burnhead, Darvel.
Steel, J., High Newton, Darvel.
Stewart, Captain Duncan, R.N., of Knockrioch, Campbeltown.
Stewart, J., Overmoor, Newmilns.
Stewart, ——, Brawlandknowes.
Stewart, Mr., Carrot, Eaglesham.
Stobo, J. & J., West Millridge.
Stone, ——, Ardochrigg.
Storry, J., Wester Crosswoodhill.
Struthers, R., Side, Strathaven.
Swann, James, Collierhall, Douglas.
Taylor, Samuel, Holmhead, Muirkirk.
Thorburn, Robert, Stonehill, Thankerton.
Tod, James, Eastercash, Strathmiglo.
Torrance of Burnfoot.
Turnbull, P. M., Smithstone, Gartley.
Veitch, ——, High Craigton.
Watson, G. L., Wheatpark, Lanark.
Watson, R., Culterallers.
Watters, Mr., Glenample, Lochearnhead, Perthshire.
Welsh, John, Balkerr, Castle Kennedy.

Welsh, T., Earlshaugh, Representatives of the late.
White, William, Nesbit.
Whyte, ——, Spott.
Willison, John, Acharn, Killin.
Wilson, J., Crosswoodburn.
Wilson, David, Carbeth, Killearn.
Wood, ——, Freeland.
Woodrop, W. A., of Garvald House, Dolphinton.
Young, John, Rauchan, Biggar.

The Knowhead flock, which is referred to above and elsewhere in this book, was one of the best and oldest in the country, until its dispersion on May 13th, 1886, on the occasion of Messrs. Foyer quitting that farm. The flock was established in the second half of last century, by Mr. David Dun, a former tenant of the Kirton or eastern portion of the present farm of Knowhead, which includes the famous Campsie Glen. This Mr. Dun is described in the old statistical accounts as the "Bakewell" of Scotland, in recognition of his efforts to improve the blackfaced breed of sheep. He met his death in 1794, by falling off the Campsie burn bridge when leading a wether across it; and Mr. Foyer, the great-great-grandfather of the late tenants, who already occupied the western portion of the farm of Knowhead, as it now is, took over Mr. Dun's land and the sheep stock, the latter at valuation. From then till the date of its dispersion last year the Knowhead flock ranked as one of the purest in the country, and many of the best stocks of the present day are largely indebted for blood procured from that source. The Knowhead sheep won many honours at the Highland Society's and other shows. The sale taking place at a time of great agricultural depression, the prices realised were not what might have been expected, but considering the eminence of the flock, we think it well to put the prices on record here.

Sale List of Knowhead Flock, May 13, 1886.

		Average price.			Total.		
700	Ewes and lambs	£2	0	10	£1429	10	0
38	Ewes in lamb	2	0	6	76	19	0
111	Eild ewes	1	7	8	153	10	0
120	Two-year-old wethers . . .	1	4	7	147	10	0
270	Ewe hoggs	1	11	9	428	10	0
100	Second hoggs	1	5	6	127	10	0
16	Third hoggs	0	15	0	12	0	0
113	Wether hoggs	0	19	0	107	7	0
10	Aged tups (highest price £12 10s.)	5	3	6	51	15	0
72	Shearling tups (highest price £24)	4	5	8	308	9	0
6	Ewe hoggs	3	2	6	18	15	0
15	Ewe lambs	3	1	6	46	2	6
20	Shott ewes and lambs . . .	1	1	0	21	0	0
1591		£1	16	9¾	£2928	17	6

It is worth noting that Messrs. Foyer sat out their lease of the farm of Knowhead at a rent of 13s. per sheep, while the farm is now let to another at a rent of about 9s. 4½d. per sheep, of which rent £50 is to be spent annually in permanent improvements on the farm. Mr. M'Indoe, the new tenant, it should be added, secured about two-thirds of Messrs. Foyer's stock at the sale.

CHAPTER II.

CHARACTERISTICS OF THE BREED.

IN giving the characteristics of blackfaced sheep it will be interesting, in the first place, to quote the descriptions given of the breed by writers of an earlier period. The true blackfaced sheep is thus described in "Walker's Hebrides," vol. ii. :—

"His body is of a plump barrel shape, his head is horned, and his face and slender legs are as black as jet, without any mixture of white. His face is set off with a thick prominent collar of wool surrounding the neck. He is the boldest, the most hardy and active of all the sheep kind. He fattens readily, and to a considerable size. When this is the case, and when he is of a proper age with access to heather (heath), his meat by general consent is preferable to every other sort of mutton, that of the small native race perhaps excepted."

Naismyth, who made a tour through the sheep pastures in the southern parts of Scotland, 1795, writes :—"In the Lammermoor district sometimes a fallow or eild ewe from the hill, killed, weighs from 9 to 10 lbs. per quarter." With reference to wool the same writer says :—"Eight hogg fleeces, nine ewe fleeces, and six wedder fleeces make a stone of 24 lbs. avoirdupois. The length of the staple is from 4 to 5 inches long."

"In the Lanarkshire district," Naismyth says, "from six to seven fleeces make a stone ; the wool is not washed before shearing. Smearing is general, and in the central parts of the county the tar is very grossly laid on, with very little mixture of butter. The lambs are mostly white, but some have black spots on different parts of the body, and one perhaps in thirty-six is black all over."

" In Peeblesshire their greatest defects are the quality of the wool, and not being disposed to fatten at an early age. The fleece is often marked with blue or grey spots. The length of the staple is from 4 to 5 inches."—*Naismyth,* 1795.

Provost Johnstone, Sanquhar, writing to Mr. Naismyth at this time, referring to Dumfriesshire, says :—

" The sheep are of the blackfaced Scotch breed. As to their origin it is difficult to say; they have been in this county time immemorial. Few of them have been mixed with other kinds. Many farmers have been in use to buy in rams from different places (though still of the same kind) for improving the breed, and this being attended with good effects, has become a general practice.

" Originally the blackfaces were short-woolled, ill furnished in the fore-quarters, and small-sized, which defects have been considerably removed by the change of rams above mentioned.

" The best kinds will not exceed in weight 12 lbs. per quarter, with about 6 lbs. tallow, from the feeding they meet with where bred; when carried into superior pasture wedders of four or five years old will rise to 15 or 16 lbs. per quarter.

" Their mutton is delicious, and at perfection at the above ages (five or six years); the farmers, however, who keep wedders of whom there are but few, sell them at three years old."

Sir John Sinclair, in his " General Report of Scotland," 1814, says :—" The produce of these sheep (blackfaces) in mutton and wool is different, according to the quality of their pastures. Their flesh is well known to be excellent, and their fleece very coarse and of little value. The average weight of a fat wether is stated at about 52 lbs. the four quarters, and of a ewe 40 lbs. In Tweeddale or Peeblesshire, which may be considered the headquarters of the breed in Scotland, the prices in 1813 were, for fat lambs, from 10s. to 14s. Ewes in lamb in spring, from 23s. to 24s.; when fattened on turnips the following winter, 24s. to 26s. Wethers three years old, for feeding, 27s. to 29s.; when fed on turnips, from 32s. to 35s. Their wool sold in 1811 and 1812 at 10s. per stone of 24 lbs. avoirdupois, and in 1813 at 12s.; but afterwards, in the course of the summer, was worth 17s. From six to seven fleeces make a stone of 24 lbs. These prices, however, are above the average of stocks and seasons. By comparing a variety of communications, it appears that the medium price of white (unsmeared) wool, for a number of years past, may be about 8s. 8d. per stone of 16 lbs. of 23 ounces, or 4½d. per lb. avoirdupois; that of smeared wool is reckoned at 25 per cent. lower. The sheep of all ages, on an average of different stocks, are also fully 10 per cent. below the prices quoted; in some of the higher parts of Galloway at least 20 per cent."

Professor Low, in his work on "Domestic Animals of the British Islands," 1845, says:—"This breed (blackfaced) possesses characters which distinguish it from every other breed in the British Isles. It is of the smaller races of sheep, with respect to the weight at which it arrives; but is larger and more robust than the Zetland, the Welsh, and the ancient soft-woolled sheep, which it displaced. It somewhat resembles the Wallachian; and, as the latter has an affinity with the Persian, it might be conjectured that it is derived from the East. But it is more natural to assume that its peculiar characters have been communicated to it by the effects of food and climate, in the rough heathy districts from which it is derived. The male and female have horns;

very large and spirally twisted in the male, but sometimes disappearing in the female. The limbs are lengthy and muscular, and the general form is robust; but the shoulders are not so low as in the Welsh breeds, nor are the posterior limbs so long. The face and legs are black, and there is a tendency to this colour in the fleece, but there is no tendency to the brown or russet colour which distinguishes the older fine-woolled races. The fur is shaggy and the wool coarse, in which respect it differs from that of all other mountain breeds in the country. It is of medium length, and weighs about three lbs. the fleece when washed. The ordinary weight of the wethers, when killed at the age of about four years, is fifteen lbs. the quarter, but some are made to exceed this weight, when properly treated and sufficiently fed from an early age. The mutton is not so delicate as that of the Welsh sheep or the South Downs; but it is more juicy and has more of the venison flavour, and is preferred to every other by those who are used to it. The mothers are hardy nurses, and are able to bring up their young when they themselves have been exposed to severe privations. In snowy weather this wild and hardy breed will dig up the snowy surface to reach the herbs beneath, and support life under circumstances in which the more delicate races would perish. They do not, like the sheep of Wales, prefer the summits of the mountains, but feed wherever pasturage can be obtained, and are not so nice in the choice of herbage as other races, derived from countries yielding finer grasses; and although wild and independent in their habits, they are not so restless as the mountain sheep of Wales and other parts, but can be induced to remain in enclosures, when sufficient food is supplied to them."

Mr. Henry Stephens, in "The Book of the Farm," a more recent work, says :—"The blackfaced ram has always an arched nose, expressive of boldness and courage. The face and lips are covered with black hair, or mottled with white, mostly the latter. The head is horned; and the horns, being large and curved in the aged tup, are considered the most picturesque objects of their kind exhibited by any animal of this country. The wool is somewhat long and coarse, which renders it of comparatively small value as an article of manufacture, and, being rather thin-set, exposes the body to the inclemency of the weather. To assist the animal to withstand the weather, the fleece is subjected to the filthy operation of smearing, which deteriorates the value considerably. The carcass is well formed, carrying its depth forward to the brisket better than the Cheviot; but still the entire body is narrow, owing to the flatness of the ribs, which renders it light—or in want of *substance*, as it is commonly termed. The flesh is fine-grained, high-flavoured, greatly esteemed, and can be sufficiently fat on the turnips and pastures of the low country. The breed is very hardy, frequenting the highest parts of our heath-clad mountains, and in summer require little care from the shepherd."

The prominent features of a good specimen of the breed, at the present time, are a rough, shaggy fleece, great agility, hardy appearance, and a bold, defiant action. Both ewes and rams are horned. A model ram has the following characteristics :—Head large and masculine; Roman nose, with wide-open nostrils and black muzzle; the face covered with a variegated black and white, or sometimes all black, close, hard hair, the colours clearly defined and not running into each other. The black, as a rule, predominates, although many well-bred specimens have rather most white on their faces. Large full eye, and broad between eyes. Horns strong and nicely curved clear of the side of the head, and about an inch apart at the roots—never meet on the cantle, nor rising above the level of the cantle. Tups whose horns rise much above the crown of the head are objected to for breeding purposes, as the horns in the lambs are

B

liable to cause severe injury to the ewe at lambing.* The
neck is rather short, strong, and slightly crested. Shoulder
level and well filled up to the neck. Back straight and not
too long. Ribs well sprung and deep, giving the animal a
round barrel-like appearance. Back broad. Hind-quarters
deep and fleshy. Deep and broad chest, with wide brisket.
Strong legs, especially from the knee upwards. Large feet,
with open hoofs, and springy pasterns. Such a shaped foot
is a great safeguard against foot-rot, as a close-hoofed foot
cannot be so easily cleaned and cured, and is not so
answerable for climbing steep hills. The wool is strong
and thick in staple, and about 12 to 22 inches in
length; slightly "wavy" and free from hairs and dark or
blue-grey spots. Blue or black streaks about the neck
or tail-head, though still common to the breed, are not
desirable. The wool, when full-grown, reaches to within
an inch or two of the ground. The legs are generally of a
jet black colour, and squarely planted under the body. The
chest and hind-quarters are broad and square, imparting
fine symmetry to the frame. The movements are elastic and
active; and a ram with all these properties is majestic, and
carries himself with great style.

The blackfaced ewe is in all respects similar to the ram,
only more feminine in appearance, and having much weaker
horns. As a rule, the colour of the face shows more dark
than white; some being entirely black. The horns in ewes
should also spring low and wide at the root, and be entirely
free from a reddish tinge (blood-horned), otherwise the
animal may be regarded as soft or unhealthy.

In point of prolificacy as breeders, blackfaced ewes are
about equal with our other varieties of sheep. In their
native condition, on healthy farms, about 95 lambs can be

* It is desirable that ewes should be bred with wide hind- (gigot)
quarters, so as to enable them to give birth to large lambs, and useful
for crossing purposes.

calculated on from every 100 ewes. But, as may be imagined, the character of the weather, both at the tupping season and at lambing, has a great deal to do in determining the percentage of lambs reared. In a bad season the number of lambs may be reduced to 60 or 70 per cent.; while in exceptionally favourable seasons a full percentage is sometimes reared. Many twins are not wanted among hill sheep of any kind. The quality of the grazings are such that the ewes are too poorly fed to supply sufficient milk for a pair of lambs. If enough twins are produced to fill the place of those that perish, the hill farmer desires no more, and is well satisfied with from 90 to 95 per cent. of lambs.

A very remarkable characteristic of the breed is the activity of the new-born lambs. With lowland sheep the lambs, after being dropped, take a considerable time before getting on their pins, but it is not so with blackfaces. After a shake of the head and a look around, the youngsters are on to their feet and sucking in less than five minutes. But for this quality, arriving as they do often in the midst of snow, many of them would freeze to the ground before regaining consciousness. When the lamb is newly born, if well bred, every inch of it except the hoofs is thickly covered with wool. It is this characteristic which renders the breed so valuable for high, exposed grazings, and it is also a quality which distinguishes the blackfaces among other breeds when they are reared in lower situations.

Nature and Habits.—Blackfaced sheep are peculiarly adapted for enduring the hardships and privations of life, which they are of necessity compelled to encounter. They are, by nature, less dainty in their food than any other kind of the sheep species. Pasture that will maintain blackfaces in good condition is often so scant and poor that the wonder is, even to experienced people, how ever they manage to exist. In many parts of Scotland the greater portion of their food consists chiefly of heather, with only a blade of grass intermixed at wide intervals. Like other animals, they prefer better food when it can be had; but their indomitable energy when occasion requires in picking up a bare subsistence from comparatively barren hills places them in this respect far ahead of all other races—the Welsh and Herdwicks not even excepted. Their remarkable action and powers of endurance enable them to climb steep and rugged mountains, and to travel long distances with comparative ease in search of their daily needs. In the winter season their lot is seldom one either of peace or of plenty. On the contrary, it is more often one of storm and starvation, against which they struggle at the expense of bodily comfort. In hurricane of wind and driving rain they bravely breast the mountain brows, and in deep snows will busily scrape and dig with undaunted courage to obtain the merest support to life.

"It is believed," says Naismyth, "by most farmers that these sheep can live on harder fare and shift better for themselves in bad weather than the finer-woolled ones, and that the same pasture will maintain a greater number of them."

Sir John Sinclair writes :—" The food of these sheep, summer and winter, is the same. They very rarely get hay or any other food than what their pastures afford. It is surprising how hardy they are, and how little injured, even when the snow lies several weeks. They dig and scrape for the withered herbage, and face the driving storm with much resolution when all other domesticated animals seek shelter from the care of man."

As a rule, too, they acquire a strong attachment to certain parts of their grazing, which they visit with unerring regularity every day in the year. When fairly well attended by a shepherd, the flocks are accustomed to come together in the evening to rest overnight on the highest and driest portion of their grazing, from which point they start off in the morning to all parts. The same members frequent the same haunts daily; and should the shepherd observe any irregularity in their movements, such as their feeding on the low ground later in the afternoon, or starting earlier in the morning than usual, he is warned that a change in the weather is soon likely to occur. Most wild animals are endowed with the instinct of being able to foretell meteorological events, when it is observed they feed more greedily than when the weather is in a settled state. In this respect blackfaced sheep are singularly acute, and their display of storm-signals has on many occasions been of great service to the shepherd in enabling him to take precautions for their safety in deceptive and ill-to-judge weather.

Their attachment to "home" is also sometimes strongly evinced when they have been sold and removed to other farms. Even should they be taken from the barren moor and transferred to the richest of lowland pasture, if unconfined, they will immediately seek to return. As it is, in spite of fences and other barriers, they frequently manage to return to their native walks, and that too from incredible distances. Some of them that have acquired in their youth the habit of jumping walls have been known to travel fifty miles homewards, across country intersected with both fences and deep rivers. When fairly imbued with the desire to return home no obstacle is too great for them to overcome, and, travelling night and day, the time they take to accomplish the journey is often marvellously short. On high-lying farms, where it is customary to remove the flock to lower grazings in winter, some of the older animals have been known to leave their summering regions and travel many miles to the wintering grounds, as soon as they felt the first signs of snow; and, in a like manner, if fine weather set in before the usual time of removal from the lowlands, the shepherds have great difficulty in preventing some of them from escaping in the direction of the summer grazing. At the lambing period, also, they evince a singular attachment to certain localities. They usually select the same spot of ground to lamb upon year after year. Sometimes they wander to secluded parts, where doubtless they imagine no harm will come to them; but the reverse is more likely to occur if left unseen, and it is therefore necessary to herd them in close quarters at that period of the year.

BLACKFACED RAM, "SEVENTY-TWO."

(Illustrations, pp. 16 and 22.)

DESCRIPTION.

First prize shearling at H. & A. S. Show, Inverness, 1883; first prize two-shear at H. & A. S. Centenary Show, 1884; first for Best Ram, any age, at H. & A. S. Centenary Show, 1884; first in family group at H. & A. S. Centenary Show, 1884; sire of first prize Ram Lamb, H. & A. S. Centenary Show, 1884; sire of first prize Lambs in male family group at H. & A. S. Centenary Show, 1884; first prize Aged Ram at H. & A. S. Show, Aberdeen, 1885; sire of fourth prize shearling Ram at H. & A. S. Show, Aberdeen, 1885; sire of third prize Ram in aged class at H. & A. S. Show, Dumfries, 1886; sire of fourth prize shearling Ram at H. & A. S. Show, Dumfries, 1886. Sire, "Seventy-one;" g.-g.-sire, "Young Afton;" g.-g.-g-sire, "Afton;" g.-g.-g.-g.-sire, a Ram that gained first prize when ten years old at Sanquhar Show, and purchased when a Lamb from Mr. M'Kersie, Glenbuck, by Mr. William Sharp for Mr. Dryfe, late of Barr. Dam, a Ewe of the "Black Diamond" foundation. (Bred by "Glenbuck." No. 18 in Stud Book.)

CHAPTER III.

DISTRIBUTION AND NUMBERS.

TWENTY years ago the distribution of the blackfaced breed could have been traced on the map with a considerable degree of accuracy; but at the present time such a process would be useless and impossible. The breed has extended into nearly every district of Scotland, and no county could now be named that does not contain considerable numbers of them within its borders. They have also penetrated through the counties of Northumberland, Cumberland, Westmoreland, Lancashire, Yorkshire, and Derbyshire, in England, and into Wales, where they are now to be found in great numbers. As long as the blackfaces remained in a comparatively unimproved state, it was seldom that people living in lowland districts saw anything of them. They were strictly confined to the highest and wildest of the mountain ranges, and lowland farmers regarded them as unfit to tread upon the same quality of land along with the Cheviots or Leicesters. But all this has changed. The blackfaces, which were formerly looked upon with contempt, proved their worth on the highest-lying grazings in the country, and gradually they began to encroach upon the ground held by the Cheviots. They had been steadily improving in size and quality, which also fitted them for lower ranges, and every year a few more flocks were added to their strength, till at length they have become the most numerous and widely distributed breed in Great Britain.

In their march towards the lower pastures of the country they have, in many instances, fairly ousted the Cheviots from old established quarters, and at the present moment they are contending with lowland breeds for supremacy on the very choicest quality of land. And in this they have succeeded to a surprising extent. It is now quite common to see large flocks of blackfaces depasturing on the richest and best farms in Scotland, and when given a similar chance, they are found not only equal to, but superior to most of the lowland breeds.

The numbers of the various existing breeds of live stock in Great Britain have never been definitely ascertained. The Government returns only give the total of all breeds combined. How much more instructive these returns would be if they gave the numbers of each breed separately ! We could then also see how they were distributed in each county, and notice from year to year their rise and fall in popular estimation. In the absence of such data, and being anxious to arrive at a tolerably correct estimate as to the position of our several breeds of sheep, the following table has been prepared according to the nearest approximate information available. The area of hill and arable pasture, as well as the total number of sheep given for each county, is from the Agricultural Returns of Great Britain for 1886. The estimated number of the different breeds, while not presuming positive accuracy, is perhaps not far from the true mark, if that could be definitely ascertained. At any rate the figures have not been set down at random, but are the sober judgment of capable authorities in each county, and are as follows :—

County.	Area of Pasture (Acres).		Number of Sheep in each Cy.	Estimated Numbers.		
	Hill.	Arable.		Blackfaced.	Cheviots and Half-breds.	Leicesters, Shrops, and Cheviots.
Aberdeen . .	645,711	445,771	153,637	92,814	14,412	46,411
Argyll . . .	1,967,091	66,871	959,798	932,612	23,760	3,426
Ayr	410,499	193,991	313,275	216,060	34,542	62,663
Banff . . .	243,543	59,412	53,078	41,990	6,263	4,825
Berwick . .	99,581	89,897	268,482	49,871	93,564	125,047
Bute . . .	115,237	13,420	44,008	30,746	11,156	2,016
Caithness . .	338,961	40,218	104,483	29,964	67,104	7,975
Clackmannan	15,564	6,899	8,802	1,326	6,427	1,049
Dumbarton .	121,508	25,124	66,132	21,251	19,257	25,630
Dumfries . .	443,746	130,909	460,754	239,896	192,116	28,742
Edinburgh .	93,401	66,535	159,092	51,130	43,463	64,526
Elgin . . .	202,211	36,440	53,644	31,216	17,321	5,107
Fife . . .	64,522	91,386	79,355	13,504	27,960	37,891
Forfar . . .	308,945	85,908	127,434	61,981	41,017	24,436
Haddington .	56,417	39,238	122,653	16,723	42,637	59,360
Inverness . .	2,558,267	72,674	615,722	539,546	61,213	14,945
Kincardine .	124,768	39,627	31,236	16,429	9,422	5,385
Kinross . .	17,521	18,319	28,235	4,726	11,826	11,673
Kirkcudbright	397,178	118,137	345,476	283,725	42,379	19,372
Lanark . .	313,759	146,397	203,390	173,434	18,976	10,980
Linlithgow .	17,872	27,995	17,031	7,463	8,121	1,447
Nairn . . .	99,878	9,124	15,147	7,198	3,729	4,220
Peebles . .	185,197	23,095	182,473	89,774	45,420	47,279
Perth . . .	1,309,797	150,261	673,955	497,865	93,540	72,550
Renfrew . .	64,869	52,691	30,431	16,329	7,542	6,560
Ross and Cromarty	1,909,784	43,460	298,883	160,230	126,157	12,496
Roxburgh .	243,745	93,443	452,482	39,416	362,516	50,550
Selkirk . .	143,155	13,317	148,538	44,598	82,630	21,310
Stirling . .	179,808	57,763	109,897	72,210	16,973	17,714
Sutherland .	1,306,975	11,848	211,825	67,460	136,741	7,624
Wigtown . .	166,381	83,324	109,003	82,313	17,459	9,231
Totals . .	13,659,954	1,363,193	6,348,710	4,280,784	1,285,596	782,330

It will be observed that the table only includes the mainland counties of Scotland. In the adjoining islands many thousands of the blackfaced breed exist, besides what prevail in the northern counties of England. Fully one-third of the total sheep-stock in the north of England are black-faces. In Ireland, too, they occupy a large tract of country, and even in Canada and the United States are becoming popular and numerous. Owing to their wide and scattered distribution, it has not been thought advisable to attempt to estimate those outside the mainland of Scotland. It is quite certain, however, that the blackfaced breed are by far the most numerous in North Britain; and indications are not wanting to prove that they may ultimately become the most extensively farmed breed in the world.

Assuming that our estimate is substantially correct, some interesting deductions may be drawn therefrom. In another chapter it is estimated that the capital required to stock a sheep-farm amounts to £1, 15s. 8½d. per sheep; so that, from the given total numbers, it follows that the aggregate capital invested in blackfaces in Scotland alone amounts to no less a sum than £7,634,065. Again, supposing the Leicesters, Shrops, &c., pay on an average a rent of 12s. 6d. per sheep, the Cheviots and half-breds 7s. 6d., and the black faces 5s. 6d. per sheep, the total rentals of each breed amount to as under :—

Blackfaces.	Cheviots and Half-breds.	Leicesters, Shrops, &c.
£1,177,215, 12s.	£482,098, 10s.	£488,956, 5s.

The blackfaces are, therefore, incomparably the largest rent-paying breed we have of the sheep kind, or any other animal. Their individual value is a trifle less than that of some of the others, but for numbers and total rent-production they dwarf the combined strength of all the rest, and are without doubt the leading sheep of the day. Notwithstanding that they are still, so to speak, a young and growing race, they hold the premier position; and it is only within recent years that they have begun to assert their superiority. For want of admirers and faith in their rent-paying qualities, the breeding of blackfaces has been much neglected. Prejudice and lack of experiment have been the two great barriers against which they have had to contend; but now that these have been partially removed, they increase in popularity every day. In high, bleak situations no other breed can approach them as profitable stock; and when tested against others in the lowlands, on better food and under better care, the good class of them, as a rent-paying animal, are superior in every way, but for a slight difference in the value of their wool. Their mutton is superior to all other breeds, and, from the amount of food consumed, they yield a larger quantity of meat, which brings the top price in the markets. All this may sound very much like puff, but every word of it is strictly true, and is borne out by actual facts, which any one who takes the trouble to investigate will not fail to discover.

The principal markets for the sale and purchase of black-faced sheep are held at Lanark, Perth, and Inverness. They are also numerously represented at such places as Oban, Ayr, Peebles, Lockerbie, Hawick, and Rothbury, and are to be found more or less at every market in Scotland. The best descriptions are disposed of at Lanark, the land in that district being evidently well adapted for rearing a superior class of sheep. In the north of Scotland the sheep are a shade smaller, owing to the nature of the soil upon which they are reared; but generally they are quite as well worth buying to take to other districts. The same attention has not been paid to their breeding as in the Lowlands. When the change agrees with them they grow better than the bigger sheep of the south, and even old ewes sometimes increase one-third of their original weight, while their wool grows again, like that of young sheep. Breeding flocks are chiefly kept in the southern half of the country, the wether lambs from the same being transported to the Highlands, where, after grazing for one or two years, they are again resold to feeders in the Lowlands. In the north, ewe or wether flocks are kept according to the altitude of situation. The best selection of aged wethers were to be had at north-country markets in past years, while the southern marts are famous for draft ewes and lambs. The wether trade is diminishing, owing to deer forests, and old wether (five years old) mutton not paying.

Blackfaced ram sales are held annually in the month of September at Edinburgh, Lanark, Perth, Oban, Ayr, and Inverness. At these centres most of the ram-breeders are represented by large consignments, principally shearlings, but during the last few years many lamb rams have also been sold. The Duke of Argyle, Ballymenach, and Mr. Howatson of Glenbuck each hold a special annual sale of rams at their respective farms, where young sires and ewe lambs of the highest individual merit and breeding can always be obtained. Over 5000 blackfaced rams, including ram lambs, are sold annually in Scotland at an average of about £5 each. The highest-priced ram last year sold at £75. Further particulars of rams and ram-breeding are given in another chapter.

CHAPTER IV.

BREEDING AND SELECTION.

IF the intelligence of man could by any possibility define certain causes which would insure certain results in reference to the procreation of animals, infallible rules might be laid down for the guidance of the inexperienced. But this cannot be absolutely accomplished; therefore the breeding of all kinds of domestic animals must, in some measure, remain subservient to the decrees of nature. Still there are many obvious principles which so generally lead to similar consequences, that they may be very faithfully held as acknowledged facts. Nature appears sometimes to thwart man's hopes with little incidents which, at the first glance, are not only unexpected, but seem to be unaccountable; yet, upon reflection, the mystery can often be satisfactorily accounted for. It is to control, so far as lies in human nature, such results as may be most conducive to profit and convenience that experience and circumspection are so important to the successful management of a breeding flock.

One great object in breeding domestic animals, whether sheep, horses, or cattle, has been to improve as well as perpetuate the respective species; and upon this reasoning we may safely infer that the present status of our flocks and herds, by a gradual process of improvement, now greatly excel those of former ages. The aim of the flock-master is to breed that class of animals which will pay best for the trouble and cost of production. To accomplish that object

he selects and cultivates those animals which are most likely
to attain the greatest weight and maturity in the shortest
time and on the least quantity of food.   It is more important in the case of blackfaced sheep to have a close thick
fleece to protect the animal—if you get less for the wool you
get more for the sheep.   Besides size and early maturity of
carcass, the wool-bearing properties of sheep have also to
be kept in view.   Both of these qualities will occupy the
breeder's attention—the first, in every case, as his grand
purpose ; the second as valuable, but regarded more as a
subsidiary consideration.

To attain these objects, he examines his flock carefully,
and in the course of his observations he perceives that some
members of the flock are in better condition than others.
There is the same attention paid to all, but the profit is
presumably much greater in some than from the majority
of their companions.   Being anxious to account for this, he
compares the good with the bad, and he observes that there
is an evident difference of conformation, in fineness of bone,
compactness of form, substance, symmetry, and quality.
He studies these points, and he finds that there is more or
less of this variation in conformation of every sheep that
perceptibly outstrips its fellows as profitable stock.   He
inquires further, and if he has employed different rams, he
perceives the one that possesses most perfectly this peculiarity of form and its accompanying aptitude to fatten, was
the parent of these promising sheep, or their dam had these
points in considerable perfection.   He now begins to form
some notion of the kind of animal that the profitable sheep
should be ; and he has living proof that these valuable properties may and will descend to the offspring.

The improvement of our domestic animals has been
accomplished by breeding in accordance with a few funda-
mental principles.   Sometimes these have not been understood by those who acted in accordance with them.

Oftentimes improvement has been retarded by failure to recognise them. There is still much in relation to the art of breeding which is not understood; but, fortunately, the essentials to success are known and comparatively easily applied in practice.

The characteristics which an animal possesses at birth it has obtained from its parents. What it is to become, the extent to which its powers are to be developed, will depend very largely on its surroundings—its food and care, the climate, the help or hindrance it has from man—but it can possess no quality which, somehow, it did not receive from or through its parents.

The most casual observation shows that, as a rule, with comparatively few exceptions, the offspring resemble the parents; that like produces like. This is the one great foundation principle from which the whole science and art of breeding start. Without it there could be no certainty, no hope of any permanent improvement. This is called the law of inheritance or heredity. Any characteristic of the parent may be transmitted to the offspring, whether it be an outward, visible quality, as colour or shape, or one relating to the disposition, or some internal and invisible quality; whether it be desirable or undesirable, of much or little importance. It is not true that a quality possessed by the parent will always appear in the offspring, but it may. The distinction between the words *may* and *will* should be kept in mind in this discussion.

No two animals are exactly alike. Each animal has two parents. These have not precisely the same qualities; hence the offspring cannot exactly resemble both. Here we have one cause of variation. Experience shows that the offspring may closely resemble one parent, or it may be intermediate between the parents in the qualities in which they are unlike. We cannot foretell with certainty which parent will impress its distinctive characteristics on the off-

spring, or whether there will be an intermediate manifesta-
tion.   The rule of practice from this principle of heredity is
obvious.   Select animals as parents which possess the characteristics desired in the offspring.   First of all, the successful breeder must know what he wants—must have a definite
idea of the qualities he desires.   Then he will select animals
possessing these as nearly as possible.   If he can· find but
one animal with the desired characteristic, he may find this
reproduced; if both parents possess it, the probabilities are
great that it will be reproduced.   If any animal, otherwise
desirable, possesses one objectionable quality, the wise
breeder will endeavour to avoid any approach to this defect
in the other parent.

An animal may receive from one of its parents a tendency
which does not appear.   There may be a tendency to some
disease of which favourable surroundings prevent the development, or there may be a possibility of development
in a certain direction; but circumstances may never cause
this to be manifested.   Many sheep have lived and died
without this fact being suspected; and so of any other
quality.   It will readily be seen that this undeveloped
quality may descend to the offspring and appear in it, perhaps, because of more favourable conditions.   Experience
shows us that this often happens.   Frequently an animal
does not closely resemble either parent in some point, but
does have a close resemblance in this to a grandparent, or
some still more remote ancestor.   In such cases it is most
natural and sensible to say the animal inherited this quality
through its parent, although the parent did not apparently
possess it, the characteristic having descended from the
more remote ancestor.

While, in the large majority of cases, animals closely
resemble one or both parents, the instances in which they
more closely resemble a more remote ancestor, in some one
or more characteristics, are not infrequent.   Hence we

modify the statement of the foundation principle, so that it reads: Like produces like, or the likeness of some ancestor; the offspring resembles the parent, or some more remote ancestor. This modification is called the law of reversion or ativeism, or "breeding back." Occasional instances of striking unlikeness to either near or remote ancestors are hard to explain. Monstrosities, abnormally or imperfectly developed animals, are produced, and we cannot certainly tell why. We have what we call "freaks of nature," or "sports;" but the number of these is comparatively small. The breeder may rely with great confidence on the almost certain resemblance of all the progeny of his flocks either to the near or remote ancestry.

The rule of practice from all this is, again, easily seen. The wise breeder will endeavour to select as breeding stock animals which not only possess the characteristics he desires in the offspring, but which have descended from ancestors also possessing them. Experience shows that the possibility of the reappearance of any quality rapidly decreases as the number of generations since it was seen increases. If both parents have any quality in common, probably the offspring will also have it. If the parents and all four grandparents have it in common, the probability of its transmission is greatly increased. If all the ancestors for five or six generations have had this characteristic, its descent may be expected with almost absolute certainty, and it is not important whether we say it is transmitted from the near or remote ancestor.

Good and bad qualities, it is to be remembered, are alike hereditary. It is just as important, therefore, to have a well-bred ram to begin with. The ewe, as a rule, is more precocious than the ram. She comes earlier to maturity, shows more aptitude to fatten, and has a finer quality of mutton. On the other hand, the ram has a larger frame and heavier fleece. In crossing with two pure-blooded

C

animals we secure all these advantages in their full ; whereas, with a mongrel ewe we are apt to lose in the offspring here precocity and quality, which we want quite as much as the large frame and heavier fleece of the ram. It is from ignoring this law, we suspect, that the results of cross-breeding are so often found unsatisfactory. By inattention to it, even in the first cross, although we may get a large-framed, heavy-woolled sheep, there is no certainty that it will prove a kindly or profitable feeder.

Too much care cannot be exercised in selecting rams of reliable blood and breeding, and the longer their pedigree is, the more certain will they be to transmit whatever qualities they possess to their progeny. Thoroughly healthy and vigorous animals should only be chosen for breeding purposes. Breeding for disease may be described as using an unsound animal for breeding purposes, or subjecting pregnant animals to cruelty in any form. Too much care cannot be taken to avoid all conditions which will injure or excite pregnant animals, as such conditions have a serious effect upon their offspring. Sometimes a crippled animal may be used for breeding purposes, when it has been crippled accidentally ; but when the trouble arises from disease, especially if hereditary, the animal should be discarded entirely for such purposes.

The true rule is to keep the best always for breeding. Weed out the poor animals, and sell them for what they will bring. Get rid of them, for you cannot afford to keep them. This practice persisted in, you will, after a time, find fewer culls among your flock—a large proportion, indeed, conforming to the standard of the best. And this statement may be emphasized to those who are trying to produce stock to be again used for breeding purposes. While we expect general excellence and uniformity, every breeder knows that there will be occasionally an inferior animal even in the best flocks. To the breeder who has paid long prices for his animals, and who is anxious to realise from them at once, it may seem a hard alternative to weed out and sell for ordinary prices animals that he had hoped would realise good returns. But it will pay; and in no other way can the best qualities be perpetuated in the flock, or its credit be maintained in the eyes of the public. In selecting rams aim to secure those with the best pedigree, that have symmetrical forms, good constitutions, and have the quality of early maturity. Sheep that combine these points will not fail to be profitable, if they are given only a reasonable opportunity.

CHAPTER V.

PEDIGREE—AND A FLOCK BOOK.

"That 'Blood will tell' all thoughtful men agree ;
  But whether good or bad the story be
  Which thus is told, depends entirely
  Upon the blood itself—its quality.
  If bad the blood, the story bad will be ;
  If good the blood, a story good we see."

AMONG breeders of blackfaced sheep, the question of pedigree has only been slightly studied. Leading flockmasters, however, attach much importance to its value; but, strange to say, there is only one breeder—Mr. Howatson of Glenbuck—who can produce anything like an authenticated and intelligible register of the breeding of his stock. It no doubt requires great trouble to post up a record of the breeding of all the lambs bred on a large farm, and the difficulty of keeping each strain of blood separate and distinct is unquestionably no small undertaking; but when the object is accomplished the advantage gained should go far to compensate for the labour and expense involved.

A great many breeders of blackfaced sheep have what they call a private index to the breeding of their favourite tribes; but it should be clear to the most casual observer that a much better method would be to register their flocks, and that by the institution of a public flock-book. That the time is ripe for such a proposal being carried into effect is evident from the great number of these private registers in existence, which also indicates that most breeders recognise the value of pedigree. It is true that few of them can give authentic proof of the breeding of their most valuable sheep; but it is nevertheless a fact that the breeding of these animals is well known to their owners, and the only reason why their written pedigrees cannot be produced is because the means have not been taken to record them with intelligence and accuracy. Now this is much to be regretted, not only because the buyer is unable to derive the same benefit he would otherwise be able to secure in disposing of the progeny, but because the breeder fails, in the first instance, to receive the remuneration to which he is entitled for the care and direction of his own work in producing such animals. If a pedigree is to be any service to the owner or maker of it, it must be put before the public in a clear and reliable manner; and in no other way can that object be attained so cheaply or so well as by the institution of a public flock-book.

The value which is now put upon the best specimens of the blackfaced breed would also indicate that their pedigrees should be preserved. No farmer can afford to pay high prices for rams whose produce he cannot sell with some guarantee of their breeding; and how high prices can ever be maintained without such a warranty is beyond our comprehension. Individual merit is the first requisite in every breeding animal, but it is only when it is supported by a pedigree that such stock can become really valuable. However, let us consider what pedigree really implies. Its thorough comprehension may go far to induce breeders of blackfaced sheep to see the advantages of establishing a flock-book, wherein they might publicly register the breeding of their most valuable stock.

A pedigree is the genealogy of an animal. As usually understood, it consists of the names of the generations. Its value, however, consists not so much in the number of generations through which the ancestry can be traced to some distinguished progenitor, as in the quality or character of the ancestry; and in proportion as we approach the top of a pedigree—that is, the immediate progenitors of a given animal—the more important does the character of the

ancestry become.   It is a well-settled fact in breeding that,
as a rule, the longer the line of descent in unbroken succession through ancestors uniformly distinguished for unusual
excellence, the greater is the probability of that peculiar
excellence being transmitted; hence the true value of a
pedigree consists not so much in its length as in the
merits of the individuals that compose it.   Four or five
"top crosses" with animals of rare merit make a pedigree
of much greater value to the practical breeder than ten,
twenty, or more of animals of no special excellence.

The further back this genealogy of good animals extends,
and the more uniform the quality of the ancestry, the better;
but the more immediate the ancestry in any given case the
more important does its quality become.

No pedigree can be a good one that does not usually
produce good animals, and no pedigree should be prized
above others unless it produces better animals than the
others.   If, tried by this test, any pedigree fails, no matter
how much it may have been idolised, its value is fictitious
and its effect is hurtful rather than beneficial.   The only
true aristocracy of blood is one that brings superior merit;
without this it is a delusion and a snare.   No matter what
it may have been eight or ten generations ago, if from a
wrong system of breeding, if from a lack of care in selection, if from incestuous breeding, or from any other cause
whatever, any particular strain has ceased to be uniformly
superior in itself, it has lost its patent of nobility.

The best pedigrees come from beginning with the best;
and they are maintained by rigidly excluding whatever is not
of first quality.   The disastrous consequences of beginning
wrong in matters pertaining to pedigree are unfortunately
too manifest and plentiful to need stating.

It requires a spirit of thoroughness and true devotion to
breed well and make great pedigrees.   We sometimes read
of men investing largely in pedigreed animals without
seeing them : it is inconceivable that such can have the

essential genius of the true breeder. The improvement of a flock is never the work of such men. The animal in the hands of the true breeder is as so much clay. It is something that can be made to change its type in obedience to a clear conception and a steady purpose. In this work all the higher faculties may have amplest play. Reason and observation, fancy and fact, firmness and gentleness go hand in hand. Greed of gain is a sign of unfitness; for the temptation to part with the best will surely come, and, the love of gain prevailing, we have the fall. The man who would succeed as a breeder must necessarily retain his best, year by year improving his pedigrees, and so keeping well under the reversion tendency possessed by all domesticated animals. He who in such a spirit puts his hand to the plough and never looks back is on the high road to the amplest reward.

As the most powerful means to promote valuable pedigree—*i.e.*, pedigree with individual merit—and to make that pedigree saleable, we strongly urge the formation of a Blackfaced Sheep Breeders' Society, and the publication under its auspices of a flock or stud book common to the breed. There are numerous other matters relative to blackfaced sheep and blackfaced sheep-farming which would properly engage the attention of such a society.

BLACKFACED RAM, "GLENBUCK YET."

DESCRIPTION.

First prize shearling Ram at H. & A. S. Show, Stirling, 1881; third prize two-shear at H. & A. S. Show, Glasgow, 1882; sire of the first and second prize Lambs at H. & A. S. Show, Glasgow, 1882; sire of the first and second prize Lambs in family group at H. & A. S. Show, Glasgow, 1882; sire of the second and third prize shearling Rams at H. & A. S. Show, Inverness, 1883; sire and g.-sire of the one and two-shear Rams in first prize male family group at H. & A. S. Centenary Show, 1884. Sire, "Niddrie;" g.-sire, "Benhar." Sire of dam, "Black Diamond the First;" g.-sire, "Listonshields;" dam, a ewe hogg, by "Old Grandfather," by "Bowley;" g.-dam, a prize ewe, bred by Captain Kennedy of Finnart, in 1863. "Listonshields," a prize Sheep at many shows, and bought for £50 in 1872 from Mr. Aitken. (Bred by "Glenbuck." No. 1 in Stud Book.)

CHAPTER VI.

BLACKFACED CROSSES.

BLACKFACED ewes are extensively crossed with some of the
heavier sheep of the Lowlands. In their native walks only
rams of the same breed are used, but on semi-arable and
pastoral farms which carry a stock of blackfaced ewes, Leicester rams, or those of some other breed, are crossed with
them, the produce being a valuable class of sheep, which
are known in Scotland by the name of "crosses," or "grey-
faces," and in England by the term "mules." Blackfaced
ewes drafted from the hills at five and six years of age are
also all purchased for crossing purposes by Lowland farmers,
who, after putting them to a Leicester or Shropshire ram,
fatten both the ewes and lambs for the butcher.

The extent to which these systems of crossing are practised
must be on a very wide scale, and consequently occupies a
prominent place in the sheep-farming industry of the country. There is first of all the semi-arable farms carrying
regular breeding flocks from which cross-lambs are bred,
and these, it may be affirmed, amount to no trifling numbers. Then if we inquire what becomes of all the blackfaced ewes regularly drafted from the hills, amounting as
they do to nearly a million annually, and scattered throughout the Lowlands of Scotland and England, it will be
apparent that a very large number of "crosses" are produced every year.

The first cross between the blackfaced ewe and Leicester

ram are sheep deservedly in good repute. The lambs are
very hardy, and such apt fatteners that they often go direct
to the butcher from hill pasture without having once tasted
extraneous food of any kind. A good authority in England
writes :—"It is only within the last few years that we our-
selves have proved how valuable and profitable a flock the
ewes are for breeding fat lambs. Crossed with a Border
Leicester ram, some of the finest lambs are produced, and
the ewes are such excellent mothers that we have drawn
lambs off to market at four months old weighing 15 lbs.
per quarter, and this weight is not much exceeded by the
mothers. So good is the quality of the meat that the fat
lambs command the first price in the market." As shearlings these cross-bred lambs also find a ready sale for
fattening purposes, and are preferred before even pure
Leicesters, half-breds, or almost any other breed, because
of their apt tendency to fatten on a moderately rich diet,
and the subsequent high price they bring when ready for
killing. Their mutton is superior to the pure Leicester,
and it is also better "mixed" and not so greasy as that of
the Leicester Cheviot cross; and being finer flavoured and
more palatable, the *epicures* are willing to pay a higher price
for it. The wool of the blackfaced "crosses" compares
favourably both in quantity and quality with half-bred wool,
and brings about the same price in the market.

The tops of the "greyfaced" ewe lambs are much sought
after, and make valuable ewes for breeding. Generally a
white-faced ram is put with them, and the progeny are called
"second-crosses," or "three-parts-bred." They are usually
larger than the first cross, and have imbibed more of the
whitefaced type. Ewes of the second cross are, again, in
many cases bred to whitefaced rams, and this produce will
have still further advanced towards the whitefaced breed,
possessing mutton not so prime as that from the first cross,
but more weight of it, and wool somewhat improved in

quantity and in value by the pound. These blackfaced and whitefaced crosses, generally, are constitutionally strong; under fairly good circumstances they thrive and do well, and they appear to be good rent-paying sheep on a large portion of the second-quality land throughout Scotland and the north of England. They are considered more hardy than the half-bred, and adapted to higher grazings.

The Border Leicester ram has been found pre-eminently superior to all others in crossing with black-faced ewes. Shropshires and other short-woolled breeds have been tried, but they do not give the same satisfaction. The principle by which we ought to be guided in crossing is, that the sire should be possessed of the greatest attainable aptitude to fatten rapidly, combined with a sound constitution. These qualities the Border Leicesters possess in a high degree. On the other hand, blackfaced ewes carry mutton of a very fine quality, but light in point of quantity. It is, therefore, necessary to mate them with heavy-muttoned, so that the produce may possess both bulk and a good marketable quality of meat. And when we consider that the black-faces with fine mutton are comparatively thin-fleshed, the propriety of crossing them with the thick-fleshed, ready-feeding Leicester is very apparent. On this principle, it is evident that the Shropshires, though possessed of finer mutton, but being thinner fleshed and slower feeders, are therefore not so suitable for crossing with mountain ewes. The Cotswold or Lincoln ram would rank next to the Border Leicester for putting to blackfaced ewes; any long-woolled breed, in fact, is preferable to the short-woolled varieties. Short and long-woolled breeds never cross satis-factorily, and it is owing to this reason, more than any other, that the Cheviot and blackfaced breeds have never been suc-cessfully amalgamated. It has sometimes been attempted, but experience has shown that the progeny is inferior both in shape and quality, and the practice is not to be commended.

The laws of cross-breeding act in direct arithmetical ratio. Thus the first cross is always half-and-half; but the produce of the second generation will show 100 parts of blood, on the *out-and-out* line only 25 parts of the original blood, and on the *in-and-in* side 75 parts of the blood of the ram. This divergence goes on, ever widening throughout each successive cross, until, in the tenth generation, there is only 1.1024 part of the original blood, and 1023.1024 parts of the blood of the ram in the stock. Suppose, for example, it is wished to give an infusion of Leicester blood to a blackfaced flock. If well-bred blackfaced ewes are selected and put to a pure Leicester ram, the first cross will be half Leicester and half blackfaced blood. Then if the half-bred ewe lambs are, in turn, crossed with a Leicester ram, the produce will be three-quarter Leicester and quarter black-faced; and so on each time the cross is repeated, until, in a few generations, the blackfaced blood is merely nominal, and the flock is practically pure Leicester. The black-faced strain could only be retained by frequent returns to the original blood. Similarly, after the first cross, if pure blackfaced rams were used amongst the half-bred ewes and their produce the Leicester blood would die out, unless again and again freshly introduced.

CHAPTER VII.

BLACKFACES V. CHEVIOTS.

A HARDY constitution is one of the chief characteristics of the blackfaced breed, and this qualification renders them peculiarly valuable for high mountain grazings. But for them, it may be said, the hills of Scotland would be almost worthless and entirely void of stock the greater part of the year. They thrive and prosper in situations where the Cheviots cannot subsist, and they can also be successfully bred at a much higher elevation. Where Cheviot wedders are grazed—and that is a step higher in the mountains than they can be bred—blackfaced ewe flocks can be kept with profit all the year round; and wedder flocks of the latter breed can exist in situations where the former would perish outright. A blackfaced sheep will thrive almost anywhere and in any climate, if it can only secure a vestige of herbage of any kind. Their powers of endurance are truly extra-ordinary. On good keep or on bad, they seem equal to the occasion; and where they cannot be maintained at a profit on high ground, no other breed of sheep in existence will prove more remunerative.

It is sometimes asserted that on certain hill lands, not by any means the highest-lying, blackfaces are not so suitable as Cheviots, owing to the conformation of the soil and pasture, which for some reason or other does not agree with them. Such instances, however, are comparatively rare, and as far as we can learn, the reason may be found in putting on stock which has been bred on an entirely diffe-rent kind of soil. If the blackfaces are bred on the farm, they will thrive as well as the Cheviots; but they should not be judged by the performances of stock brought from a strange locality. This is merely a complaint common to all breeds, and blackfaces are not excepted.

Blackfaces are well adapted for nearly every description of soil in Britain, more so than any other breed that could be named. The Cheviots were hardy, good sheep at one time, but during the last twenty years they have deteriorated in the former respect very considerably. They have been bred too big and soft for being suitable to any but the very best quality of hill grazings, and are little different in con-stitution from some of the Lowland breeds. In regard to quality of wool, the Cheviots have always had the pull over the blackfaces; but that is their only superior point, and it, too, seems to be on the wane. Only a short time ago, a Hawick tweed manufacturer was declaiming that Cheviot wool was now of so coarse a quality that it was unsuited for the purpose it was formerly used, and in consequence was bringing a lower price in the market. It is not many years since the price of Cheviot wool was reckoned at double that of blackfaced, but at the present time the difference is only about one-third more; and since wool of any kind became of so little value, the difference on the aggregate is a mere trifle.

In respect of mutton qualities, the blackfaces are un-doubtedly superior to the Cheviots. In former days, when the Cheviots were in their glory, all the eild ewes became prime fat during the summer months, and could be sold to the butcher in the autumn direct from their hill grazing. This is never experienced now-a-days. The eild sheep in the flock are little better-conditioned in autumn than those that have reared a lamb, and they have to be sold in the same way, to be subsequently fattened off on cake or corn.

On the other hand, blackfaced ewes that run eild come off the hills in autumn thick fat, and fetch a relatively higher price. This is a very remarkable fact, and one of more consequence than many suppose. If the eild sheep do not fatten, how will it be with those bringing lambs? The results are identical. Again, thousands upon thousands of Cheviot lambs were at one time bought by butchers at the autumn lamb sales for the London fat market, but now scarcely a single lot is fit for this purpose. Blackfaced lambs have taken their place; and since the breed has been improved they bring as good prices as the Cheviots did formerly. In recent years the seasons have been so unfavourable that few hill lambs of any kind have been fit for killing, but what proportion have been suitable for this purpose were nearly all blackfaces. "The Cheviots are out of the field," was the remark an extensive buyer of fat lambs made to us recently; "they want the tallow for killing."

Again, with blackfaced wedder flocks the improvement which has taken place in recent years has been no less remarkable. The idea has generally prevailed that wedders of this breed could only be profitably fattened after they had attained to three or four years of age. Such an opinion is now only held by those who have no recent experience of the breed. On exposed grazings it may yet be advisable to graze wedders until three years old, but for all practical purposes they can be sold a year sooner, and feeders are as ready to buy them at two years of age as at three. To the grazier a year's shorter keep makes a huge difference in the profits. Take, for example, a flock consisting of 3000 sheep: instead of selling 1000 at three and a half years, 1500 can be sold at two and a half; and the latter, if well bred, will bring about the same price per head in the sale-ring.

Blackfaced wedder lambs are also now readily bought for

fattening as hoggets, and the weights they grow to at four-teen months have surprised even their most confident buyers. Taken from the hills in autumn, and put on ordinary fatten-ing fare in the Lowlands, they easily attain a weight of 20 lbs. per quarter, and produce a quality of meat that is not excelled by the Southdowns, while it brings an equally high price in the market. This is a trade which has sprung up quite recently, and is still unknown to a great many sheep-feeders.

There is another field where the blackfaces are deservedly popular. This is on semi-arable and pastoral farms. On such farms it has long been the custom to keep a stock of Cheviot ewes, from which a half-bred lamb was taken. But Cheviot ewes cannot bring half-bred lambs without being turnip-fed in winter, and we have no hesitation in saying that this practice has been the ruination of many farmers in Scotland. When a Cheviot ewe is turnip-fed in winter, and turned out to graze on hill land in summer, the lamb she rears is seldom a good one, and the price it brings is seldom equivalent to the cost of production. Fortunately this system of sheep management is no longer necessary. A blackfaced ewe grazed on the same pasture, can rear a cross-bred lamb, which will bring quite as high a price in the market as the other, without the aid of a single turnip in winter. This makes a wonderful difference in favour of the blackfaces; and if old ewes of this breed are of less value individually than Che-viots, they are not more difficult to fatten, and cost a pro-portionately lower price to begin with, which all goes to prove their superiority over the Cheviots.

One more point in favour of the blackfaces calls for com-ment, and that is the shepherd's opinion in the matter. Ask any man who has herded both Cheviots and Blackfaces which of the two breeds he would rather tend, and his pre-ference will invariably be in favour of the latter. It is in spring and at lambing-time that the stamina and constitution of hill sheep are put to the test. In bad seasons the former are often so weak and spiritless that to turn them with a dog is as much as their life is worth, while the latter are strong and hearty, although existing under the very same conditions. But the greatest difference, so far as the shep· herd is concerned, is experienced at lambing-time. The blackfaced lambs are more easily satisfied with milk, and that is a very great advantage where the ewe's opportunities for milk-production are so meagre. On the other hand, the Cheviots, being leaner in condition, give a less quantity of milk and of a poorer quality, while their lambs require comparatively more; and the contrast on this account is really a very striking one. At lambing-time the blackfaces require only half as much attention and nursing as the Cheviots. A shepherd who has once had experience with a blackfaced flock will never again herd Cheviots if he can possibly avoid it; and as an instance of this kind, where necessity compelled it, an over-anxious shepherd of our acquaintance, who removed from a blackfaced hirsel to a Cheviot one, found himself overwhelmed with so many diffi· culties at lambing-time, that he went out of his mind and had to be taken to a lunatic asylum.

It cannot but be interesting to store farmers to hear what a former proprietor of Crossflat, the late Mr. Robert Aird, had to say on the question of Blackfaces *v.* Cheviots at the beginning of the century. "Our sheep," he is reported as saying, "are all of the blackfaced kind, and very handsome. Some years ago, Admiral Stewart introduced the Cheviot breed, which he kept up for some years with no profit. As soon as the land was let, the tenant hastened to get quit of them, as every tenant will lay on the stock from which he can best pay his rent. The Cheviot sheep's only property is the fleece, which is certainly good; but, then, they never fatten so well as the blackfaced sheep. They are lank in the quarters. They do not feed well on our coarse grass. When the lambs are dropped from their mothers they have little or no wool upon them, and on our wet ground they die fast. In snow the Cheviot sheep must be fed by the hand, for they are bad *workers*. They stand together in crowds and cannot be forced out, when the blackfaced sheep are spreading wide, and working in the snow to the shoulders almost. Induced by the high price of Cheviot wool, several farmers, at very great prices, have stocked high ground in the upper part of Lanarkshire, to their ruin almost. For these two years past they have done very badly, some farmers having scarcely one in ten of their hoggs living, and they are again getting back the blackfaced kind. The Cheviot breed are subject to many disorders, particularly the worst of all, the *scab*. Our wool, though not the finest, is not the coarsest. I have sometimes sold mine at £5, 10s. the pack of thirteen stones, Ayrshire tron-weight." At that time the average price of a blackfaced sheep was about 3s. 6d. per head !

A correspondent of the *Ayr Observer* writes :—

"The important question between the comparative pro-fitableness of blackfaced and Cheviot sheep for hill pasture is very much one of circumstances. On grazing land, well drained and sheltered, and in comparatively good seasons, the Cheviots would probably yield the greater profit. But on the occurrence of such severe seasons as were expe-rienced in 1860, in the immediately succeeding years, and in the spring of 1876, not only does death decimate the flock, but the constitutions of those which survive are so impaired that they never entirely recover their original vigour. . . .

" The blackfaces are a peculiarly active and hardy race ; they can endure hunger and cold to an almost incredible degree ; they are excellent workers, being adepts at digging for their food in deep snow; they are very superior as milkers, and, moreover, have a strong maternal instinct,

D

which prompts them to stick to their young offspring, even under the most suffocating drift.

" But apart altogether from the character of the seasons, there are large tracts of exposed, mossy, heath pasture for which the blackfaces are best adapted. Unquestionably a decided reaction has set in in their favour in recent years, so much so, that, instead of the Cheviots making further encroachments upon them, not a few stocks of the latter variety are brought, by gradual crossing and otherwise, to be blackfaced. This reaction has been helped by the exceptionally high prices which can now be got for ewe lambs of that breed, a large number of which are now used for rearing cross lambs by Leicester and Lincoln sires."

CHAPTER VIII.

BLACKFACED MUTTON AND EARLY MATURITY.

THE title of this chapter may sound not a little strange to the ears of breeders of blackfaces, as the idea has generally prevailed that early maturity in mountain sheep was simply a qualification not to be desired. The reason of this, no doubt, was owing to wool being of more consequence than mutton. So long as graziers were receiving from 2s. to 3s. per lb. for wool, the production of mutton was a secondary consideration, but now the reverse is the case. Wool has become the secondary consideration, and mutton the primary; and there are good reasons for thinking that this disposition of affairs has assumed a permanent condition. On that account blackfaced sheep must more than ever depend upon their value as mutton producers; and since the production of mutton is the chief object in all our breeds, it may be useful to inquire what constitutes real merit in a mutton sheep.

As producers of a fine quality of mutton, the blackfaces have no superior. Their meat has a peculiarly delicate flavour, which is much prized at the tables of the rich. Around the mansions of the nobility it is common to see a number of wedders of this breed, kept for the double purpose of ornament to the parks and supplying meat for the household. The usual age at which blackfaced wedders were formerly considered fit for killing was three to five years; but now so greatly has the breed been improved in size and early maturity, that lambs sold from the grazings

in the month of August at 14s. to 16s. per head to feeders
are again resold by them in June (then fourteen months
old) to butchers at from 45s. to 52s. each. Such a price
could only be obtained for three, four, and five-year-old
wedders some years ago. The average weight of the wedders under two shear is about 20 lbs. per quarter, and in
more favourable conditions they can be made to scale
upwards of 25 lbs., dead weight. Old ewes which have
served for breeding purposes, after being fattened, scale
an average weight of 20 lbs. per qr. Blackfaced mutton
always brings the top price in the market, and the butchers
make a point of displaying their black points in their shop
windows, whereas the whitefaced sheep are generally dis-
played minus the head and feet.

Now what is the first point of excellence one naturally
looks for in a mutton sheep? Is it quantity or quality?
The poor man would put quantity first, and the rich man
quality; so that on the very threshold of the inquiry we are
met with apparently irreconcilable differences. Between
the two extremes, however, there is an intermediate point
which perhaps combines the highest possible excellence in
both of these respects, and this, it will be found, is generally
the most valuable meat. Consequently, the sheep which
yields the greatest quantity of a given quality of mutton
must be, on the whole, the most profitable animal, when
the conditions for its production are in keeping with the
results. But, since the circumstances and conditions which
govern the distribution of the breeds vary so much, it is
impossible that every one can grow the mutton *par excellence.*
And well it is so perhaps, because we have the various
classes of mutton suitable to the requirements of all. Yet
breeders should remember, whatever be their opportunities,
that to aim for quantity without quality, or quality without
quantity, is a deviation from sound principle, and likely to
be the least profitable system to pursue.

The idea that prime mutton can only be had from three or four-year-old blackfaced sheep is altogether erroneous. It has been proved to the satisfaction both of feeders and consumers that prime mutton can be had from one-year-olds, and certainly at much less cost than from animals two or three years older. As a rule the bulk of the mutton of the present day is decidedly superior to that of the last generation, which fact may be attributed to better bred animals, and their arriving earlier at maturity. Some tell us that young mutton is neither so toothsome nor so economical as that from aged animals, but this can be disproved. Lamb is always considered a choice dish, and this being so, what reason is there to suppose that during the next twelve months mutton from the same sheep should become tasteless and unpalatable, and afterwards regain its flavour with advancing age? The truth is, the whole secret lies in the feeding. When a lamb is properly fed from its birth onwards, it is fit to kill and eat at any age. As it grows in age the flesh becomes firmer and tougher in the grain, and, consequently less easily masticated and digested, which sensations have by some been construed to mean more satisfying to the stomach. But it is an entirely false idea that the quality of the mutton deteriorates with the earlier maturity of the animal.

Blackfaced sheep have always been justly celebrated for the superior quality of their mutton. But with any breed there is often a wide difference between the mutton products of different pastures. As every farmer knows, some plants are not any more nourishing than others, but certain of them are more proper food for particular species of animals; and others again are better adapted for forming flesh and muscle than for fat producing.

Real merit in mutton sheep is, however, determined by the cost of production. Whatever may be said for or against quantity or quality, young mutton or old, unless it

can be produced at a profitable rate, it has no other grounds
for recommendation. And this is exactly where the advantages of early maturity appear to be of most service to
blackfaced sheep breeders. When sheep can be made to
weigh as much at fifteen months and to yield an equal
quality of mutton to what they previously did at three years
of age, the difference in the cost of production must be
patent to every one; and it is satisfactory to note that this
is what has been achieved by some of our more energetic
breeders within recent years. How much higher weight in
daily gain will yet be recorded it would be difficult and useless to predict, but we may rest assured that further improvement in the same direction is being earnestly striven after.
Though some breeds have here an advantage over others,
owing largely to their situation, it should not discourage
those who are now far behind in the race. It may be a
distant prospect for Scotch blackfaces to beat the South-
down record at Smithfield, but if the former can be turned
into the general market a year earlier at the same value,
the advantages are none the less apparent.

In order to encourage the feeding of *one*-shear blackfaced
wedders, Mr. Howatson in 1887 offered handsome prizes
for the same, to be competed for at the Highland and
Agricultural Society's Show at Perth; and he proposes to
repeat the offer in 1888. The competition for these prizes
was not so great at Perth as could have been wished, but
this was easily accounted for by the lateness of the announce-
ment, and few or no breeders being prepared for it; and
there is little doubt that the entries for these prizes will be
far more numerous next year. The Highland Society have
for many years been in the habit of giving prizes for blackfaced wedders not above *four*-shear; but our national
Society should be more prompt to recognise progress in
this matter of producing sheep for the requirements of the
million. Four-year-old wedders are out of date altogether,

and are no longer considered profitable stock in ordinary farm practice. If proof of this were needed it might be found in the fact that Mr. Russell's second prize pen at Perth in the class for wethers not above four-shear were only one-year old. They were purchased as lambs at Lanark last year for 13s. a head; yet in July last they weighed 20 lbs. per quarter, showing that blackfaced sheep are neither slow in maturity nor light in weight when properly cared for. Early maturity in every race of sheep must henceforth receive the greatest possible attention; and in no breed has this quality been more neglected than in the case of blackfaces. Custom and habit have led many people to believe that it was unprofitable to fatten mountain sheep before they had reached a certain age; but recent experience has shown that blackfaces can be fattened with as much success and profit under fifteen months as any other variety of sheep in existence. This is a fact which cannot be too widely known; and if feeders of young sheep will only consider that the buying-in price of blackfaced wedder lambs is very small compared with those of other breeds, the inducement to invest in them should be very tempting indeed. In reference to the feeding of one-shear wedders, Captain Stewart of Knockriock writes a most suggestive letter, which is well worth preserving for perusal by those interested in the breed. It is as follows :—

" The very generous offer made by Mr. Howatson to give prizes for the best blackfaced wedders, one shear, at the next Highland and Agricultural Society's Show, and that of 1888, will, I hope, be encouraged by wide competition in 1888 at least.

" I am sorry I have none this year feeding for the purpose; but I will put up two lots next autumn, and feed under different circumstances, and ask the Society to allow me to exhibit them and give an account of their treatment at the show of 1888; and I hope that Mr. Howatson's effort may lead to the gradual doing away with mixed stocks, and the earlier maturity of blackfaced sheep. I know of a lot of half a score now under treatment which I hope will be at the show of 1887 ; and although they were not taken in hand as early as they might have been, I am sure they will astonish those who have never seen this class of sheep fed before they are four or five shear.

"The reason I propose to feed two lots is this : I believe very high feeding, like very high farming, does not pay ; in other words, it does not pay to make a manure cart of an animal's stomach, just as it does not pay to leave manures (dung or so-called artificials) in the ground all winter, or more in the land than the crop can assimilate ; and just as it will not pay to keep a woman looking up and down an acre of turnips for the very last weed in it. I believe the most profitable way of feeding is to keep the animal in robust health, with as little exercise as may be compatible with health, and as much warmth as will be found to pay ; in fact, warmth and food are to a great extent interchangeable ; but beyond this, no more food should be given than the machine or stomach can convert into flesh or bone of equal or greater value. I will, therefore, feed one lot regardless of expense, and another in the way I consider they may best pay me.

"May I propose the coming Highland and Agricultural Society's Show as a fit time for the breeders and feeders of blackfaced sheep to form a society for the furtherance of these objects?"

This proposition of Captain Stewart's has not been acted upon, but it is good for all time, and by-and-bye we hope to see it carried into effect.

CHAPTER IX.

BLACKFACED SHEEP IN THE SHOWYARD.

THE influence of the showyard has perhaps done more to bring blackfaced sheep into prominence than all other efforts put together. By the spirit of emulation awakened in the show-yard, individual flocks were improved in the first instance, and this influence has spread to the remotest parts, where never a sheep was prepared for show, but all the same have been brought under the spell of improvement. Shows have done even more than sales to encourage and animate breeders to produce good sheep; and while many instances of improvement could be given where showing was never practised at all, there is little doubt that the stock of those who have taken a prominent part in the showyard have always been in the van of the others when the real merits of both came to be tested in the sale-ring.

Showing has sometimes been credited with the ruin of Cheviot sheep, but the same cannot be said, at least, of its effect upon blackfaces. On the contrary, the results have been highly beneficial in every respect; and so long as neither judges nor exhibitors are led astray with false notions as to the breeding of animals fit for the showyard only, good, certainly not harm, is certain to follow. Judges of show sheep have a great power in their hands for good or evil in the matter of breeding. By favouring animals of a different character than it is absolutely necessary blackfaced sheep should possess, to fit them for high grazings, the current of improvement may be diverted into what might really prove an injurious and retrograde channel. This is the mistake alleged to have been committed with reference to Cheviot sheep. Competition between rival breeders became so keen that actual merit which fitted these sheep for hill-farms was overlooked in favour of soft tendencies which belong more properly to sheep suitable for arable land.

By straight breeding even it is possible to depart from the true characteristics of the breed, and it is highly important that only sheep of the right stamp for hill grazings should be encouraged in the showyard. This involved a question not only of vital consequence, but one very difficult of solution. To the simple question, What is the proper stamp of a blackfaced sheep? very different views might be held. In the first place, opinions regarding the size of the animal would vary. Those having pasture capable of growing a large sheep would advocate a heavier animal than those whose farms could only carry a small one. But this is a point of less importance than that of hardiness of constitution, upon which the very ablest of sheep judges may sometimes be deceived in their opinions. Unless a blackfaced sheep is thoroughly *hardy*—a word which means far more than it really conveys—it is of no value whatever for mountain purposes; and the difficulty is to determine between what are termed *hardy* and *soft* specimens.

In respect to wool, there is less difficulty in arriving at a just conclusion; still, here also opinions might vary as to the kind of fleece best calculated for protecting the animal in a high, exposed, and stormy climate. The character of the fleece, however, has a very important bearing on the point in question—a fact which has frequently been too lightly studied by judges in awarding prizes in the show-yard. It is not unusual to see animals that have taken first and second prizes with wool as different in fibre as it possibly can be. The wool, of course, not being the only point upon which the merits of the animals are determined, variation in that respect cannot well be avoided; yet there is reason for thinking that the standard for wool and its extrinsic as well as intrinsic value should be more clearly defined than is generally the case. Wool can be grown of almost any quality, but there is one particular quality which is better adapted for the requirements of blackfaced sheep than the others. Let it be decided what that quality should be, and breeders will aim to cultivate it, to the benefit of the breed in general. The step taken by Mr. Howatson of Glenbuck last year (1887), in offering handsome prizes out of his own pocket, to be competed for at the Glasgow Agricultural Association's Show, and at the Highland Society's Show, for the blackfaced sheep carrying the fleece best adapted for protecting the animal in a high, exposed, and stormy climate, is a movement worthy of the very highest praise. It is the most sensible proposal in the way of offering premiums we have heard of for some time, and deserves the support and encouragement of all interested in the welfare of the breed.

In one respect the blackfaces occupy a unique position in our showyards. They are the only breed of sheep shown in their natural fleece—without any colouring of the wool, and without any attempt at clipping into shape. No artificial colouring would improve the natural hue of these sheep, and the object of every exhibitor is to show them as rough in the wool as possible. The only condition, therefore, which it is necessary for Societies to insist upon is that the sheep must be shorn bare on or after the 1st of January of the year in which they are exhibited.

So much indirect benefit has been derived from flocks taking prominent positions in the showyard that the practice of showing can hardly be given too much encouragement. Many more farmers than do ought to take part in at least local exhibitions of blackfaced sheep. It no doubt involves a certain amount of extra expense and trouble, but these are the very requisites by which improvement is attained; and, after all, any expense or labour incurred, if judiciously applied in the direction of genuine progress, will seldom prove an unprofitable investment. A few show sheep on the farm always create a keen interest in the stock as a whole, and induce the owner to exercise his best skill and energy in the way of breeding better animals. It even induces him to buy an extra good ram now and then, and the influence of a single ram is often a great power for good in the flock.

At all the principal shows of blackfaced sheep there are now classes for tup lambs, and there is a reason for this which does not as yet apply to other Scotch breeds. Blackfaced tup lambs are much more largely used for breeding purposes than any other sort in Scotland, and when they are fit for this service sooner than the others it certainly also entitles them to earlier recognition in the showyard.

At some of the shows—*e.g.*, Glasgow and Ayr—within the last year or two, prizes have been offered for single animals in the female classes. This is unquestionably a step in the right direction; for perfection will never be reached until each sheep can stand on its own merits. We would not, however, do away with the customary pens of three or more ewes, gimmers, and ewe hoggs; for there is an attraction about numbers which helps to make a show popular; but we would, at the same time, give prizes for the best single ewe, gimmer, and ewe hogg, and also a cup or medal for the best female sheep, as is now done in the tup classes. The Highland and Agricultural Society will do well to take the lead in this matter, and in other reforms which are inevitable in our present show system.

It is to be regretted that sheep or any other animals have to be high-fed for showyard purposes. Not that judicious feeding cannot be practised without injury to the stock, but because it prevents a great many, owing to the expense, from bringing out the sheep in show order. This is a difficulty which show reformers have long had to contend with, and no satisfactory solution of the question has yet been arrived at. Lean-stock shows are the latest movement in this direction, but unfed stock have little or no attraction for the general public. Moreover, sheep and cattle of all kinds are virtually bred for the express purpose of fattening, and if they are not exhibited in their best condition they do not enlighten people as to their best properties. Put a stop to feeding, and you put a stop to progress in other respects, or, to use a surgical expression, " the cure is worse than the disease."

Many practical farmers are of opinion that blackfaced sheep can be fed, if judiciously, without damage to either constitution or subsequent usefulness. This opinion is fully supported by at least two of the most successful breeders and exhibitors of blackfaced sheep in Scotland, whom we will take the liberty to quote, from a correspondence on the subject in the *Farming World*. We refer to Mr. Howatson of Glenbuck and Captain Stewart of Knockrioch, the former of whom writes :—

" I have never found the judicious feeding of sheep hurtful to their constitutions. I could give a host of evidence to prove this, by referring to rams that have gained first prizes as lambs, and continued to be prize sheep for three years at the Highland and Agricultural Society Shows, and are now being used successfully on hill pastures at the age of six and even ten years. I can also refer to ewes in the same way. Two prize ewes were sold by me last year at ages of thirteen and fifteen years respectively. They produced twin-lambs yearly (with three exceptions, when they had single lambs); and I learn they are again in lamb this season, both ewes being as healthy as sheep can be.

"I find sheep that will not stand the test of *feeding* will not certainly stand the *starvation test*, which is one of vital importance to blackfaced sheep.

"A well-bred tup lamb can hardly be injured by feeding as long as he is nursed by his mother, which is the case at the Highland Society's shows in July."

This letter Captain Stewart supports as follows :—

"I must express my very strong concurrence with the admirable letter on feeding blackfaced sheep from Mr. Howatson in your paper of 4th March.   My experience is quite the same as his, and I turn my sheep fed for the prize ring up on to my hill without anxiety, only seeing they never want a feed of hay or ensilage if they will take it.   I would not give them this if I could have my hill fenced, but it runs out to disputed land by three proprietors, and cannot be closed at that end, so I cannot stock my land as I would wish; therefore I have to feed on the hill.

"I have a sheep now that was fed for show as a lamb and shearling by Mr. M'Gibbon; did a season's work on the Mull of Kintyre, and was fed again; has done two years' work for me on my hill, and is as fresh and active as ever; and has placed me far up in our local shows with his stock.   I have bought show ewes ten and twelve years old, and taken prize lambs off them.   Over-feeding will kill any animal, but no over-fed animal will take a prize—he cracks up at once.   Sheep are kept for feeding, and ought to stand it.

"No one would think of trying to feed Mr. Howatson or myself.   We are not the kind ; we put our food into energy, not into fat.   But I will undertake to feed blackfaced yearlings to 100 lbs. of mutton, and I will start with the lambs of those that stand feeding."

The display of blackfaced sheep at the Highland and

Agricultural Society's Show at Perth, in July last, has probably never been equalled in any showyard, either as regards numbers or quality. There were no fewer than 142 pens of this breed forward—there being 24 aged tups, 63 shearling tups, 26 tup lambs, 12 pens of ewes with lambs, 11 pens of shearling ewes or gimmers, and 6 pens of wethers. These large entries were materially brought about by the offer of Jubilee prizes to the amount of £141, 10s. 0d., in addition to the Society's prizes. The Jubilee money was subscribed by about two hundred breeders of blackfaced sheep, through the medium of the *Farming World*, and the competitions in this section excited the keenest and most wide-spread interest.

CHAPTER X.

POINTS FOR JUDGING BLACKFACES.

THE judging of blackfaced sheep by a scale of points, though perhaps never practically tested in any showyard, is yet a system which many experienced breeders highly approve of. The present method of judging, conducted as it is solely on the uncertain principle of eye and touch, does not at all times prove satisfactory. The different judges not only see and feel differently, but they put different values on the various points of the animals before them. A scale of points, in a measure, obviates this confusion of ideas. It tends to bring the estimates of different men into closer relationship with one another, and assists them in fixing the proper value of each point; the whole when added together being the actual worth of the animal. The following scale fairly represents what we consider the character and relative value of the various points of a blackfaced ram :—

PERFECTION.

Head.—Large and masculine ; nose, thick and slightly arched; nostrils, expanding; muzzle, broad and black; forehead, broad; eyes, large and bright ; face, covered with close hard hair, all black, or black and white, the colours clearly defined, and not running into each other 12

Horns.—Strong and nicely curved, clear of the side of the head, about an inch apart at the roots, and not rising above the cantle 6

Neck.—Strong and slightly crested 4

Shoulder.—Narrow on top, and well filled up to neck . 5

PERFECTION.

Chest.—Deep and broad 12

Back.—Broad, level, and not too long 10

Ribs.—Well sprung and deep 7

Hind Quarters.—Wide, deep, and fleshy . . . 12

Tail.—Set on level with the back, strong, and hanging
well down the legs 2

Feet and Legs.—Feet large, with open hoof; pasterns,
moderately long and sloping; legs, strong, especially
from the knee upwards, of a black or black and
white colour, and squarely planted under the body 5

Wool.—Strong and thick in staple, 12 to 22 inches
long, slightly wavy or curly, and free from hairs
or blue-grey spots 15

Constitution, pedigree, &c. 10
_____

Total 100

Every breed of sheep has its own characteristics, which
can be defined as clearly as black and white. These, published, are a great help to young breeders, who, without this
guide, may waste years in fruitless efforts to improve their
flocks. In truth, with every one breeding according to his
fancy, having no fixed standard to aim at, we might as well
mix up all our breeds. But when we come to the showyard
we require to go a step beyond the standard. One judge
fancies a certain type of sheep, and his colleague disagrees
with him. Judge A. says, "What a splendid head and
horns No. 1 has!" B. says, "But look at the deep and
broad chest of No. 2." "But the wide, deep, and fleshing
hind quarter of No. 1," continues A.; and in return B.
draws attention to the heavy and beautiful fleece of No. 2.
The question is here put to Judge C., who should decide
the case; but he has to balance in his own mind whether a
superior head is more to be considered than a deep and
broad chest; and again, is a superb fleece more than a
set-off against unusually good hind quarters? And thus
points, without having some definite value attached to them,

E

might be compared one against the other *ad infinitum*, without ever coming to a satisfactory conclusion. The greatest difficulty judges have to encounter is in deciding between sheep of very equal merits. But there is no such thing as two sheep being exactly alike in every way, and were the scale test applied one would be found superior, although to the casual observer the decision might seem erroneous.

It is of the first importance that correct judgment should be given at shows. The whole influence is wielded by the judges, and if they place it in the wrong scale the whole labour of the year is not merely thrown away and lost—it is rendered mischievous, and had better not have been. Is it therefore better that judgments should be almost instinctive verdicts of experienced men who arrive unconsciously at a conclusion after inspection, or the simple arithmetical result of a valuation within prescribed limits, supposed to include the whole character of the subject under examination? If the latter be a possible process—and we believe it is—there can be little doubt of its superiority. But while many believe the process possible enough where it is a mere mechanical object or result that is under review, they can never make up their minds to its practicability on a live animal. They consider such cardinal points as vigour, constitution, pedigree, &c., cannot be placed under definite valuation, and that consequently, in awarding by scale of points, the verdict might be given in favour of the worst.

But this objection will not stand. If the scale of points is anything like correct, such a result as is here apprehended is impossible, and can never happen. A live sheep can be taken to pieces and adjudicated upon with as much ease as an inanimate machine; and, indeed, unless this is done, the reason of the awards cannot be made clear to those interested and looking on. There is also little doubt that unless each point is weighed against the corresponding point in the opposing animal, conflicting decisions will be the rule. A good shoulder cannot be weighed against a good back. It must be shoulder against shoulder, and back against back. This one consideration, in fact, constitutes an unanswerable argument in favour of judgment by a scale of points.

"HIGHLAND MARY," "CENTENARY CHAMPION," AND "SEVENTY-TWO" (*Illustration*, p. 67).

DESCRIPTION.

" Highland Mary."—Prize gimmer at H. & A. S. Show, Stirling, 1881 ; prize ewe at H. & A. S. Show, Glasgow, 1882 ; dam of the best ram lamb at H. & A. S. Centenary Show, 1884.

" Centenary Champion."—First prize ram lamb at H. & A. S. Centenary Show, 1884 ; third prize ram in aged class at H. & A. S. Show, Dumfries, 1886. Sire, " Seventy-Two." Dam, " Highland Mary."

" Seventy-Two."—First prize shearling at H. & A. S. Show, Inverness, 1883 ; first prize two-shear at H. & A. S. Centenary Show, 1884 ; first for best ram any age at H. & A. S. Centenary Show, 1884 ; sire of first prize ram lamb at H. & A. S. Centenary Show, 1884 ; first prize ram in aged class at H. & A. S. Show, Aberdeen, 1885.

CHAPTER XI.

BLACKFACED RAM BREEDING.

As may be inferred from the very large number of black-faced sheep throughout the country, ram-breeding occupies a· prominent place in the pastoral economy of Scotland. It is estimated that the total number of rams of this breed sold annually in North Britain exceed 5000 head, and that these are produced by about 250 breeders, located in nearly every county. A good many more farmers sell tups of their own breeding every year, but strictly speaking, the number who really ought to be designated " breeders " can almost be counted on the fingers of both hands. There are unquestionably far too many who attempt to earn the title of breeder. It invariably happens that when there appears to be money-making in any business a great number rush in to share the spoil. In sheep breeding nothing could be more disastrous to the flocks at large, and buyers of tups ought to show their sense of disapproval of this work far more strongly than they do. It is the simplest thing in the world for any man who happens to be owner of a flock of sheep to reserve a certain number of his lambs for tups, and the result is the ram sales are flooded with a great number of animals no more fit for breeding purposes than a wild boar is to mate with an improved Berkshire. The wonder is that sheep breeders are foolish enough to buy such rams for use at all. The small prices no doubt tempt many a man to buy inferior sheep, although he is perfectly well aware they are not what he would like if he could afford better. But nothing could be more suicidal to his own interests. A ram is a power for either great good or evil in the flock, and how any good can be got from a ram descended from a stock without any pretensions to improved blood is a problem requiring no elucidation.

A good-looking ram may now and then appear in a flock of unknown breeding, but that does not qualify him for a sire. The power of transmission of character to the progeny is only given to rams of established pedigree, and no other consideration ought to influence the buyer in making his selections. It is not imperative that every flock in the country should be pedigreed, but the most humble breeder may gather a few hints for his own benefit by merely reflecting upon the principles by which breeders of any description of thoroughbred stock are guided. Could any one suppose for a moment that any leading breeder of cattle or sheep would use a sire in his herd or flock without first having made certain about the antecedents of the animal in question. Money value is the last object such men think of in selecting a sire, and surely if the principle is the only safe one to follow in the case of valuable stocks, it is no less so in the meanest flock in the country. An inferior-looking ram from a flock of undoubted pedigree and breeding is worth double the price of one, which has a better appearance but with no authenticated character and breeding. Good breeding and good looks are of course both essential, but the former is by far the most important requisite of the two, and the breeder who grudges a few pounds for the sake of economy in the first instance is only throwing stones at his own head. No real and lasting benefit was ever derived from mongrel-bred sires, and until sheep-farmers generally begin to realise the necessity for buying and using rams of the very best breeding and quality only, so long will the markets continue to be stocked with poor sheep, bringing an equally poor price and begetting progeny of the same character.

It costs less to produce good rams than it does to produce bad ones—less, because a good-bred animal is always easiest kept and takes less time to fatten. Consequently, the higher average price that can be obtained the greater will be the profits. As long as ram-breeding can be carried on at a profit, no man need deny himself the privilege of pursuing the trade. But the truth is, a large number of the breeders annually lose a good deal of money in the business; and it would be far better, both for themselves and others, were they to leave it alone altogether. By calculating the cost of production we can very soon arrive at a correct opinion regarding the average price at which blackfaced ram-breeding becomes profitable. In the first place, estimate the value of the lamb at the ordinary figure received for the top wedder lambs of the same flock. It should really be a few shillings more, as the best are always reserved as tups, and this reduces the average quality of the wedders. Now, if the selling price of tup lambs exceeded that of the wedders by about 10s. a head—it is not worth while mentioning a smaller sum, as the expenses incurred always amount to a few shillings per head—there would certainly be some reason for encouragement, and the business, if not highly remunerative, would at least cause no loss. But do breeders on the aggregate receive 10s. a head more for their ram lambs than their wedders? As a matter of fact, many of them receive a good deal less. Between 500 and 600 ram lambs are sold annually at the Ayr sales, and we find that last year 524 of these made the disgraceful average of 10s. 7d. ! Had the same lambs been duly castrated and sold for a different purpose, there is little doubt but that they would have fetched at least 5s. per head more money in any sale-ring in Scotland. This is merely one illustration of how things stand, but many more of the same character could be given. Instead of being profitable, the sum of 5s. per head, not including expenses, was here actually lost through the pure folly and greed of unscrupulous persons. Eminent breeders, whose lambs bring from £3 to £6 a piece, are of course well enough paid; but they would be paid much better if certain individuals could be made to see that handing them out 5s. willingly, comes to the same thing in the end as having to part with it unwillingly, for the benefit of those who have no right to obtain it. Unless a sum of about £2 can be obtained for lamb rams, there is little use in attempting to breed them. In fact, a breeder whose lambs bring from 10s. to £1 ought to feel a trifle ashamed of his productions, and common sense should tell him that the public consider his stock unfit for the purpose he ventures to claim for them. If a ram lamb is not worth £2, it is worth nothing, or perhaps less than nothing, for an inferior sire may do an incalculable amount of damage to the flock in which he is used.

If lamb rams cannot be produced at less than £2 per head, what ought it to cost to produce shearling rams? This is a question which every breeder must answer to his own satisfaction; but speaking generally, to maintain a blackfaced ram intended for sale purposes the following year in September, after paying all expenses and leaving a margin for profit on capital and labour invested, the sum required will not be less than £2 per sheep. That, charging the lamb at £2, brings the average cost of production up to £4 per ram at eighteen months of age. Or, leave the value of the lamb out of the question, and take the cost of rearing a shearling from the commencement. The period of feeding may be put at 18 months, and in that time it will consume cake and corn at the rate of fully 1 lb. per day, less at first and more ultimately, or say 6 cwt. in all, costing about 35s. The hay for winter may be put at 20 stones, costing 6d. per

stone, which is another 10s. Then there must be young
grass for autumn feed, and the following summer's grass,
which together cannot be put at less than 15s. This brings
the cost of a shearling up to £3, for food alone after the
milk.

Cake or corn	.	.	.	.	.	£1	15	0
Hay	.	.	.	.	.	.	0	10	0
Grass	.	.	.	.	.	.	0	15	0
						£3	0	0

It is possible to rear the rams on a cheaper scale, perhaps,
but the price received will be correspondingly low. Those
who bring out their sheep in good style never estimate on
selling under £4; and should their returns not amount to
that sum, they have lost money. The crack lots, averaging
£12 and upwards, could afford to spend much more than
£4 in production, but it is questionable if they do. On
the other hand, many blackfaced shearling rams have been
averaging a trifle over and under £3 per head, which proves
that inferior rams are too abundant, besides showing very
plainly that numbers of breeders must have lost heavily in
the business. If farmers will persist in breeding rams
whether it pay them or not, then it is just this, " them
that will to Cupar, maun to Cupar," and there is the end
to it. But by every one exercising a little discretion in the
matter, and putting the knife to a good many more than is
now the case, the number of rams annually sold would not
only be greatly reduced, but the quality would be improved,
and the greatest benefit of all would be the weeding out
of inferior animals that pollute the country with inferior
produce.

Breeding from blackfaced lamb rams is a practice of
comparatively recent origin. Mr. Howatson of Glenbuck
was the first to inaugurate the sale of blackfaced tup lambs,
which he did some fifteen years ago. Since that time the

trade has immensely increased, and has been adopted by breeders in all parts of the country. It is useless, as some have been doing of late, to throw cold water on the practice and declare it to be ruinous. If farmers did not find lamb buying and using of some benefit they would shortly let it be known. The truth is, ram lambs have been giving general satisfaction, and every year there is a growing demand for them, which proves that for breeding purposes lambs have a far greater value than has yet been attributed to them.

The advantage gained in buying tup lambs is that it enables the purchaser to use and winter them as he thinks best, and in the opinion of some of the leading farmers, it is the cheapest and best way of procuring useful and healthy rams capable of getting sound stock. A well-grown tup lamb will serve as many ewes on any hill as a shearling bought at any of the sales throughout the season. Mr. Howatson tells us that from his experience as a sheep-farmer during a long period of years, he has used tup lambs with economy and advantage—taking care how they are selected and bred— with ewes on the highest portion of his pastures, which run from 1000 to 2000 feet above sea-level; and to enable them to travel easily on rough ground and during snow, he clips the wool off their tails, legs, and bellies. Many other farmers have expressed similar testimony regarding their experience with tup lambs, and with such strong evidence in their favour it is needless to predict that in future the demand for them will be largely increased.

The only reason why shearling rams are not more suitable than lambs, is because of their being over-fed, which spoils them for the purpose for which they are required. Unless the rams are well fed for a whole year and brought out in great bloom, no purchaser will look at them in the sale-ring. This is a practice which not only blackfaced sheep breeders have long tried to obviate, but breeders of almost every race of sheep or cattle have experienced the same difficulty. The point, however, rests with the buyers, and so long as a fat animal brings more money than a lean one, so long will the breeders continue to overfeed.

But it is impossible to overfeed lamb rams when they are disposed of a few weeks after they are weaned. Moreover, September is not a suitable time of the year for purchasing shearling and aged rams of the blackfaced breed. At that season their fleeces are too deceptive; therefore, as a protection to buyers, and improvement of the breed, all the shearling and aged rams should be sold in May or June, when they are in full fleece. One reason for the September sales is, that it answers better for occupiers of unhealthy land to take their tups home at that season. But surely it would be better to purchase the rams in May with their natural fleeces on, and, if the ground is unhealthy, make an arrangement with the seller to keep them until they could safely go home to their new pasture; then, if they are in high condition, there is sufficient time to reduce them and prepare them for the tupping season in November; or, better still, purchase well-grown tup lambs in September, winter them as may be thought best in the low country, and prepare them for the following season's use as you think they ought to be prepared.

Markets for the sale and purchase of blackfaced rams are held at various centres in Scotland, the principal of which take place at Edinburgh, Perth, Oban, Ayr, Lanark, Inverness, and other places of less consequence. At these places nearly all the principal breeders are represented by large consignments, but the best specimens are to be found at Lanark, which is the natural centre for this breed, and the sales at which have during the last year or two been fast taking first rank. Besides these sales, which are held in the month of September, Mr. Howatson has an annual special sale for rams and ewe lambs of his own breeding at Glenbuck. The Duke of Argyll also holds an annual sale of rams at Ballymenach, Argyllshire. At Glenbuck and Ballymenach the stock sold consists chiefly of lamb rams, and it is needless to say that these are of the very highest merit and breeding to be found in the country. Mr. Howatson's flock are all pedigreed animals, and purchasers in want of really first-class tups cannot find the same quality of sheep elsewhere in Scotland. The Glenbuck sale has been an annual institution for the last fifteen years, and it has come to be regarded as the great event of the year, where breeders from all parts of the country meet to do honour to the spirited proprietor, than whom no one has done more for the improvement and advancement of the blackfaced breed.

As far as we have been able to note, during the last three years, the total numbers of blackfaced rams and ram lambs sold, at public sales and privately, was 4952 in 1887, 3977 in 1886, and 4381 in 1885. The general average price was £4, 5s. 10d. in 1887, £3, 19s. 6d. in 1886, and £2, 15s. 4d. in 1885. In reality there was a much greater rise in the average price of 1887 than is here represented, for a vastly increased proportion of the rams sold this year were lamb rams. It will be noticed how steadily the average price has risen year by year since 1885. Blackfaced rams are generally used four seasons, so that, including both rams that have been purchased and rams that have been bred and reared on the farms where they are used, it is probable that not fewer than 40,000 blackfaced rams are put to the ewes in Scotland every year.

The highest price paid in 1887 for a blackfaced ram was £75, given by Mr. Howatson of Glenbuck, for a shearling selected from the flock of Mr. Murray, Parkhall. The highest price at any of the public sales was £52, 10s. This was given at Edinburgh by Mr. Howatson for one of the Duke of Argyll's lot. At Perth sales, one of Mr. Duncan's, of Benmore, lot was bought by the Duke of

Argyll for £50. Mr. Howatson also bought privately two shearlings at £50 each, from Mr. Woddrop, of Garvald, and Mr. M'Cracken. The highest average of the season was £12, 18s., made by the lot of Messrs. Archibald, Overshiels, at Edinburgh, the Duke of Argyll's lot averaging £12, 9s. 7d., Mr. Hamilton's, Woolford, £12, 3s. 1d., and Mr. Fleming's, Low Ploughland, £11, 4s. 6d. One of Mr. Fleming's lot made £38, one belonging to Mr. Craig, High Ploughland, made £33, another belonging to Mr. Craig £30, and one from Overshiels £30, these being the third, fourth, and sixth highest prices of the year at the public sales. The above averages were nearly equalled at Ayr, where Mr. Richmond's, Drumshang, lot made all round £11, one of them fetching £31, the fifth highest price of the year. The ram lambs of the same breeder averaged £2, 5s. 10d., one bringing as much as £10. Several of Mr. Howatson's ram lambs made £10 and over at Glenbuck sale, while thirteen of the best averaged £3, 6s. 11d., and another lot of thirty-three £2, 19s. 7½d. These high figures were topped by the Duke of Argyll at Edinburgh, whose average for a small lot of fourteen ram lambs was £4, 4s. 3d., the best making £11.

The following is a summary of sales and prices of the last three years :—

		No. sold.	Average price.			Total value.		
1887	Blackfaced rams .	3622	£5	4	3	£18,879	5	0
	Blackfaced ram lambs .	1330	1	15	6	2,483	15	0
	Total . .	4592	£4	5	10	£21,362	15	0
1886	. . . . .	3977	3	19	6	14,708	11	6
1885	. . . . .	4381	2	15	4	12,124	0	10

Tupping.—The rams are put out to the hill about Martinmas, or the 22d of November, or a week later in higher districts. One ram may have fifty ewes, or even a few more; but to prevent him having to travel too far, the shepherd ought to throw the ewes together night and morning. With Lowland flocks the breast of the ram is rubbed with ochre, and a different colour used every week, so that the shepherd, who makes a note of the colour used, may have a key, as it were, to the order in which the ewes will lamb; but this practice is never adopted by hill-shepherds. Early lambs are of little good on the hills, and the endeavour should be to have the lambs arrive with the grass in spring, but not before. Late lambs, however, are quite as bad as those that arrive too early, and seldom come to anything; the rams, therefore, should not be left with the ewes after the new-year at latest.

During the tupping season the shepherd must be constant in his attendance, and, if the weather permits, keep the ewes on the higher ground in detached bodies here and there, according to the nature of the ground and condition of the pasture. Each body of ewes should have a proper proportion of rams along with it, and the shepherd must be careful not to allow the rams to leave the ewes to which they are appointed. This mode of procedure appears preferable to any other, and a smaller number of rams will then suffice.

Should the weather become stormy, and the snow interfere with the feeding of the flock, the rams will be the first to suffer, and means must be taken to supply them, at least, with some extra food daily. The object in feeding rams for service is not to fatten them, but to make them strong animals, perfectly developed, with every part of the system well nourished. For this purpose there is nothing equal to a mixture of crushed oats and peas, in addition to grass, or in winter a little hay and roots. More or less extra feeding will need to be given the rams, and continued throughout the winter, after taking them from the ewes.

CHAPTER XII.

WINTERING BLACKFACED SHEEP.

THE successful wintering of the flock taxes the skill of the flock-master quite as much as the summer management. This may readily be imagined when the whole bearings of the question are considered, and more especially the variable character of the seasons. Were all our winters of the same degree of severity, it might be possible for each farmer to lay down fixed rules, and to adopt a definite system for the management of his flock year after year, but owing to the changeableness of the weather, this course is impracticable. No two winters are exactly alike in every respect. The period of the year, or the month during which severe weather may occur, or the length of time it may prevail, cannot be reckoned, and, therefore, alterations in the methods of management must be made accordingly. Preparation and promptitude in adopting such means as are most beneficial to the welfare of the flocks during any emergency of weather constitutes the most skilful management, whereas, by neglecting to provide for them such food as may be required, and delay in affording the necessary relief when the sheep actually begin to suffer, denotes the unskilful and unsuccessful manager.

Whether our winters are now less severe than formerly is a question which no one, perhaps, is very well qualified to answer. It is generally supposed, however, that the winters have become considerably modified during the present century, and that we do not experience such heavy snowfalls or such intense frosts as those recorded in former times. What reasons may ·be assigned to justify such a belief we dare hardly venture to state ; but on the evidence of the oldest living inhabitants who positively declare the winters are milder since they can remember, and from fewer losses now recorded amongst farm stock, we are forced to grant a certain amount of credit to the truth of these assertions. It may be that the accounts given of severe winters in past ages were somewhat exaggerated, but it is no doubt true, as the " Ettrick Shepherd " has remarked, that storms constitute the various eras of the pastoral life, they are the red lines in the shepherd's manual—the remembrancer of years and ages that are past, the tablets of memory by which the ages of his children, the times of his ancestors, and the rise and downfall of families are invariably ascertained.

Turning to the effects of severe winters upon hill farmers, the greatest loss of stock recorded in Scotland was during " the thirteen· drifty days of March," some particulars of which, gathered from traditions, were given by James Hogg, with, perhaps, a little poetic embellishment. The date is variously stated as 1620 and 1674, but a curious circumstance points to the latter as being correct. It is recorded that in 1675 Monmouth, then the husband of Lady Ann Scott of Buccleuch, got a licence to import from Ireland 4800 nolt of a year old and 200 horses, to make up the loss sustained on the Buccleuch lands by " the thirteen drifty days." The Sheriff of Roxburghshire, W. Scott of Minto, was cautioner that the number should not be exceeded ; but as some of the oxen were more than a year old, the Sheriff was fined £200 sterling. The facts of that remarkable storm, as related by Napier, may be briefly stated. The ground was covered with frozen snow when the drift began, and for thirteen days and nights it never ceased.

The cold was intense to a degree never before remembered. About the fifth and sixth days of the storm the young sheep began to fall into a torpid state, and all that were so affected in the evening died over night. The intensity of the frost would often cut them off instantaneously. About the ninth or tenth day the shepherds began to build up huge semi-circular walls of their dead, in order to afford some shelter to the remainder. About the same time the sheep were observed to tear and eat each other's wool. When the drift ceased there was on many a high-lying farm not a living sheep to be seen, and about nine-tenths of all the sheep in the south of Scotland were destroyed. On Eskdalemuir, which sustains 20,000 sheep, only forty young wedders were left on one farm and five old ewes on another. The farm of Phawhope, at the head of Ettrick, remained twenty years without a tenant, after which it was let at the annual rent of "a grey coat and a pair of hose." On Bowerhope, a farm in Yarrow, all that remained of 900 sheep was one black ewe, which some idle dogs chased into St. Mary's Loch, where she was drowned. "But of all the storms," says Hogg, "which Scotland ever witnessed, there are none of them that can once be compared with the memorable 24th January 1794, which fell with such peculiar violence on that division of the south of Scotland that lies between Crawford Muir and the Border. In these bounds there were seventeen shepherds perished, and upwards of thirty carried home insensible, who afterwards recovered; but the number of sheep that perished far outwent any possibility of calculation. One farmer alone lost seventy-two scores, and many others from thirty to forty scores each." On the beds of the Esk, when the flood after the storm had subsided, there were found 1840 sheep, nine black cattle, three horses, two men, one woman, forty-five dogs and 180 hares, besides a number of meaner animals."

No such severe loses have occurred amongst hill flocks

during this century, owing more, perhaps, to farmers making better provision for their sheep in winter than to the absence of severe weather. The century opened badly, for in the spring of 1800, says a writer, hard weather continued till the end of May. In the winter of 1801–2 a severe snowstorm occurred, which necessitated the removal of the sheep from the hills *en masse.* The deaths were not so great as had sometimes been experienced on former occasions, but a great deal of expense was incurred in providing food for the sheep. The winter 1815–16 was much worse than any since 1794, and the loss of sheep was greater, many farms not having lambs sufficient to make up the loss of old sheep. The year 1826 and the spring of 1827 were very stormy. There was a heavy snowstorm in April at the time of Fastern's E'en Fair, which was stopped because of the storm. One shepherd, James Scott, at Merelees, Tima, Ettrick, lost his life at the end of that month. In 1837 the early part of the winter was good till about the new year 1838, when snow came on, and there was continuous frost till the middle of April. St. Mary's Loch was frozen over for thirteen weeks, and the sheep from Ettrick and Yarrow heads were all taken to Annandale for grass. In Whamphray parish alone, about 2000 scores of hill sheep were maintained. The winter of 1854 was stormy, but 1860 was one of the very worst on record. The storm began on the 26th of October 1859, and continued with great severity all through the winter—snow, sleet, rain, and frost coming in succession. At Christmas came the most intense frost of modern times. Heavy losses resulted in the following spring, about one-fifth of the old sheep and three-fourths of the lambs having perished. In no season since the end of the seventeenth century had the loss been so heavy. On one farm, out of a flock of 2500, the loss was 440 ewes and hoggs, but instead of ninety score of lambs there were only thirty score, many of them in poor

F

condition. On another farm in Ettrick 360 sheep died; and the total loss in the county was estimated at £40,000. The winter of 1879 was severe, and for a long time sheep were fed with hay. In December 1882 the snow was deeper in the Yarrow valley than the oldest shepherds remembered, and sheep could be moved only with difficulty. A good many sheep were buried and lost under the drifts; and a shepherd perished at Whitehope in Yarrow. In January 1883 there was a heavy fall of snow, with high wind and drift, when many sheep were buried under the snow or driven into burns and drowned. The winter of 1885–86 has to be noted as one of exceptional severity to hill flocks. It was not so remarkable for intensity of frost as for continued snowfalls, which eventually gathered on the hills to such a depth as to place vegetation far out of the reach of the sheep. This led to an enormous expense in hay to keep the sheep alive, and the ewes were so greatly reduced in condition that 50 per cent. of them were afterwards unable to rear their lambs. Altogether the season of 1886 was about the worst of the century for hill farmers. The produce did not amount to more than half an average yield, the prices were bad, and rents comparatively high. The winter of 1887, and this brings us up to the present date, was fortunately of a milder character, and the crop of lambs above an average.

The intervening years not mentioned in the above account were presumably generally favourable to hill flocks, as no notice has been made of them by chroniclers of such events; but, notwithstanding, many of them were, no doubt, also disastrous to a certain extent. A very important fact worth recording in connection with recent severe winters is, that throughout Scotland the blackfaced stocks invariably suffered about 50 per cent. less than the Cheviots, and that too though confined to considerably higher grazings. On this account a tremendous flood of popularity has arisen in their favour, and they are every year fast displacing the less hardy whitefaced breed.

It must not be supposed, however, because only a few of the years have been shown to be disastrous to hill flocks, that all the others are mild and agreeable, causing neither loss nor trouble in management. Such an impression would be wholly erroneous, as there is hardly any winter, however mild, that is not severe enough to cause, to some extent, anxiety and expense. Hill sheep are naturally such hardy creatures that a storm of a week or ten days' duration is never spoken of as doing them much harm, and although the shepherd's sympathies are always with his flock, the last verse in the following poem by W. Graham, Drycleuchlee, Selkirkshire, shows that before much effort is made on their behalf in the shape of extra feeding, a good deal of patience and praying for fresh weather is first restorted to :—

> " O ewie ! wi' thy honest face,
> An' a' thy kind and welcome race,
> That's high upon the mountain placed
>   To face the storm ;
> Sae fresh wi' ilka blade o' grass,
>   Like braird o' corn.
>
> Your case at present 's very bad,
> It makes me every day feel sad,
> To see ye turned out frae the stells
>   To seek your meat ;
> An' it a' covered wi' the snaw,
>   Ay, twa feet deep.
>
> But they that can get tae the heather
> Have better heart to face the weather,
> Than those that's in the doleful tether,
>   Wi' sic a snaw,
> An' working hard a' day tegether
>   For nought ava.
>
> For a' thy roads an' paths are clad,
> An' in the hollows where ye've fed,

An' where I've seen ye mak your bed
        On summer night ;
But now ye're drawing to our sheds
        A waesome sight.

O ewies ! I am proud of thee,
Thy fleecy flocks soon may we see,
A' feeding on the grassy lea
        In fresh grey morn ;
An' ne'er again disturbed be
        Wi' sic a storm.

But we contented aye maun be,
An' wait wi' patience till we see ;
Perhaps ye'll no need trust to me,
        For Him abune
Can send a fresh an pit ye richt,
        An' vera soon."

The words of poets very often mean a great deal more, and
sometimes less, than they literally convey, and we interpret
the line "An' wait wi' patience till we see," to signify that
before commencing to feed the sheep on hay, it would be well
to wait to see if a change in the weather did not take place.
Now this is one of the most vital points in connection with
the wintering of hill sheep. When to commence to feed? is
a question that has created more discussion, more disagree-
ment, and more loss of sheep than any other in the whole
range of the subject at issue.   It is one, however, that does
not admit of any definite solution, and cannot be satisfactorily answered until the facts of the particular case are fully
known and actually witnessed. Even then experienced
men will frequently disagree. The reason for this is that
we never can tell how long a storm may last—that is, how
long the ground may remain covered with snow, which
prevents the sheep gaining access to the herbage—and
feeding the sheep for a few days only is found in practice to
be more hurtful than beneficial. When the sheep once
acquire a taste for hay or other food, they never settle after-

wards so well to their natural fare, and even in fresh weather they continue to languish for food they do not require, and which it will not pay to allow them. Rather than disturb their equanimity in this way, it is maintained, and rightly, by many good stockmen, that it is better for the sheep not to commence feeding them until it can be no longer avoided. Where the evil arises, however, is in putting off the inevitable for too long. The sheep are encouraged to struggle for a subsistence day after day, and what between hoping for fresh weather and the subsequent mischief that may arise from beginning to feed them, they are allowed to fall away seriously in condition. The point which the stockmaster has to decide is whether the loss of condition or the harm done by feeding in the event of a storm of short duration, is the greater evil. When the storm ultimately proves to be protracted, and lasts for perhaps a month or six weeks, then the earlier the feeding is commenced the better, but, on the other hand, when a thaw shortly ensues, much damage may be the result. To rightly decide what to do, is not at all times an easy matter. The character of the storm prevailing, the season of the year at which it occurs, the condition of the sheep, the food on hand, the expense involved, the benefits to be derived, and the weather prospects, have all to be put into the balance of reflection and duly weighed singly and collectively, before a decision is come to. There is one rule, however, which may be given as nearly correct as it is possible to make it under the circumstances, and it is this: when feeding becomes necessary it should be begun early, and not discontinued until there is plenty of grass.

The food most suitable for winter feeding hill sheep is, without doubt, hay, or ensilage made from the same material. With a good supply of these no serious losses of stock need occur. In good seasons, however, hill farmers will still find it most economical to convert their meadows and bogs into hay rather than ensilage. Ensilage cannot be made at a cheaper rate per ton than hay if the weather be dry; only it affords a means of securing the winter's supply of food independently of the weather, and this is a great advantage in upland and moist climates. But it must not be supposed, as has frequently been asserted, that ensilage can be made equally as well in wet weather as in dry. The truth is, that silage of the best quality is made by partially drying the grass, and the more water it contains the less nutritive it becomes.

A good deal has yet to be learned in regard to the actual value of silage as a food for hill sheep, but there is every reason to believe that it offers considerable advantages in the way of economical management. If it can be proved that silage will enable hill farmers to winter their hoggs at home, instead of having to send them to low ground, at a cost of from 6s. to 8s. per head every winter, then one of the greatest obstacles to profitable hill-farming will be overcome. Some of the largest farmers in the Western Highlands are at present experimenting in this direction, but no one as yet has come to a definite conclusion on the subject. Still, so far as these trials have gone, they point to being very satisfactory.

The reason why so few have as yet adopted this modern system of preserving their fodder, is doubtless owing to the heavy expense involved in fitting up the supposed necessary silos and weighting appliances. But it has been practically demonstrated that farmers can make capital ensilage without spending money on either buildings or costly machinery, merely by stacking the grass, and covering it with a layer of earth dug from around the stack. This method of constructing a silage stack is quite simple, and may be described as follows : Build a layer of grass to the height of about six feet, taking the precaution that in doing so the grass forming the outer edge of the stack be well trampled, and that this outer edge be kept, during the erection of the stack, a little higher than the middle of the stack, the grass being merely forked into the centre without trampling. Allow time for this mass to heat and settle down ; then throw up another layer of grass about the same thickness, giving time for this also to heat and solidify, and repeat the process as often as may be needful. When the stack has attained a suitable height, then fill in the middle of the stack with grass, treading it firm and leaving the top of the stack of a shape enough rounded to give watercast. The soil around the stack may then be thrown up so as to form a bed of soil about eighteen inches thick over the top, and the work is done, no other covering being required.

Turnips or other artificial foods are not considered suitable for maintaining hill sheep in winter. In certain circumstances, that is when no other food is available, or the sheep excessively low in condition, the use of corn has sometimes of necessity to be resorted to, but the majority of farmers admit that its effects are decidedly injurious. This may be explained in the fact that hill pasture is not sufficiently rich to afterwards maintain the sheep in a thriving condition. Next to an open winter no other food for hill sheep has been found equal to the natural hay grown upon the farm, and every opportunity should be taken to secure as much of that as possible. It is a mistake, however, to suppose that that cut from amongst the sheeps' feet is better than that grown in enclosed meadows. Where the sheep are allowed to graze they very naturally pick out all the finer grasses, and thus the quality of the hay made therefrom is reduced in quality and feeding value.

HOUSE OF GLENBUCK.

CHAPTER XIII.

A PROMINENT BREEDER.

MR. CHARLES HOWATSON of Glenbuck, to whom this volume is dedicated, as a small acknowledgment of the time and money which he has so freely expended in his untiring efforts for the improvement of the blackfaced sheep of Scotland, is the eldest son of the late Mr. William Howatson of Cronberry, Auchinleck (who was the second son of Charles Howatson of Craigdarroch and Cronberry), and of Margaret Reid, second daughter of John Reid of Duncanzinier, parish of Auchinleck. Mr. William Howatson was an enterprising farmer, and died in 1882. His son Charles very early in life exhibited a taste and liking for blackfaced sheep, and, as Mr. Howatson of Dornel, he was known as a generally successful exhibitor at the district and Highland Society's shows many years before he became the fortunate proprietor of Glenbuck.

At the age of fourteen years he left his father's house, and was two years in the office of the well-known firm of Messrs. William Baird & Co., of Gartsherrie Ironworks, Coatbridge; and afterwards was engaged out of doors in ironworks and colliery management. When Messrs. W. Baird & Co. purchased Muirkirk Ironworks, in 1856, he was trusted with the organisation and management of the new purchase, which has been a most successful undertaking up to the present time, although it was otherwise under the management of former companies. Besides continuing the management of the ironworks, Mr. Howatson in 1863 leased the farm of Crossflat, in the parish of Muirkirk; and in 1865 he took possession of the farms on his estate on Dornel; until, in 1870, he retired from the iron trade, from which time he has devoted his whole attention to farming operations.

Prior to 1863 the farm of Crossflat had been owned and farmed, for a century and a half or more, by the much-esteemed family of Aird, a member of whom Mr. Howatson married in 1859. Robert Aird, who owned and occupied the farm in 1810, is conspicuously mentioned by Aiton in his report as one who was well worthy of imitation, in having, with great profit, converted heathy wilds into productive fields, and in causing rye-grass and clover-seeds to grow on what was shortly before unproductive moorish ground. Robert Aird was succeeded in 1822 by his son Alexander, who farmed the land from that time till his death in 1863. He must have been as enterprising as his father, for he carried on and extended the improvements with great energy, planting patches of wood in well-chosen places, to provide shelter for the blackfaced flock, and fencing and liming a large extent of rough hill pasture with the best results. At his death in 1863 he was succeeded in the property by his nephew, who, having no taste for farming, offered a lease of it to Mr. Howatson of Dornel, by whom it was accepted, the rent being fixed at the rate of 10s. for each ewe and hogg kept on the farm during summer. In these times this may be considered a high rent; yet Mr. Howatson has occupied the farm ever since at that figure: the explanation being that shortly before the end of the lease his wife succeeded to the family property, the farm of Crossflat included.

The lands of Crossflat lie at an elevation which varies from 800 to 2000 feet above sea-level, the latter being the height of the conspicuous mountain Cairn Table,* which forms the boundary-line between Ayrshire and Lanarkshire. The hill pasture consists principally of heather and draw-moss.

Extensive improvements were carried out by Mr. Howatson upon the land after leasing it in 1863. Much of the hill pasture was injured by water, and it was thoroughly dried by sheep-drains. By this means the evils which are well known to result from a wet pasture ground were obviated. Not only were the quantity and quality of the herbage increased, but foot-rot and unsoundness were in a great measure banished. The system of liming, which had been pretty extensively practised by the Messrs. Aird, was also followed by Mr. Howatson; and there being limekilns upon the property of Crossflat, this was done at a comparatively small outlay in cartage.

Simultaneously with these improvements upon the land, Mr. Howatson began to improve the flock, which, when he got possession of it, was only of secondary quality. That

---

* "Cairn Table," one of the rams illustrated in the present volume, was so named from the fact of his being lambed on the top of that mountain. He realised in 1870 the handsome sum of £60—a price which was then unprecedented for a blackfaced ram, Mr. J. N. Fleming of Keil and Knockdon being the purchaser.

he did so with great enthusiasm, and at large expense, will be understood when we mention that during the first five years he spent as much money in the purchase of tups as he received during the same period for the whole of his wedder lambs. This liberality he soon found to be remunerative, as he thereby began to reap the profits of a first-class flock at a much earlier stage than would have been the case had his improvements been accomplished at less expense and at a less rapid stride.

The foundation of the improved flock, on the female side, was laid by the purchase of a pen of ewe hoggs belonging to the late Captain Kennedy of Finnarts, which were first in their class at Ayr Show in 1864. The sprittle faces of that gentleman's flock were remarkably fine in quality, their only deficiency being a want of bone; but, by a careful and judicious selection of the sires put to them, that want was readily supplied in their offspring.

Mr. Howatson's breeding from his purchased animals was in the first instance strictly experimental. The rams were put to selected ewes, and their produce were carefully scanned, in order to determine whether or not the cross had proved a successful one; and only such as came well up to the mark were used as breeders. Not a few tups which in themselves were really good-looking animals were discarded as sires after they had been proved, from a season's trial, to be disappointing in the quality of their offspring. We need scarcely say that experiments con-ducted so systematically and carefully made the process of improvement a very sure one, while the liberality in the purchase of tups, to which we have already referred, prevented its being slow.

From the very first start at Crossflat, Mr. Howatson's object has been to produce the best type of blackfaced sheep which can be produced for the climate and class of pasture on which they have to graze—a type of sheep with a compact frame on short legs; a good head, with full, prominent eyes and nostrils; and well-shaped horn. Strong nose and jaws are also points which he has steadily kept in view, as showing a high state of breeding and strong, hardy constitution. He has striven to have the sheep covered with a large quantity of well-grown wool, not only on the bodies, but also on the bellies, and on the thighs down to the very knees. In this he has succeeded so well that the newly-born lambs are woolled all over with a thick natural coat, which is so protective against the most severe weather that we have seen them skipping and frisking amongst the snow before they were half a day old. Blackfaced lambs are naturally nimble on their legs; they cannot be this, however, if the weather is severe and they are stricken down with cold; but the woolly-skinned lambs are safe in any weather. It is equally important that the old sheep should not throw off their wool, even when in low condition, before the natural season for casting the fleece; and Mr. Howatson, thinking this tendency the result of bad breeding, has striven to have his flock so bred as to be superior to this weakness, however low in condition they may become. The consequence is that, after a severe winter, even a " broken coat" in the flock is very rarely seen. The weight of the fleeces has also greatly increased since Mr. Howatson began his improvement of the flock. In 1864, which was a fairly representative season, the average weight of the fleeces of ewes and hoggs was $3\frac{1}{4}$ lbs., whereas in 1875 it was $5\frac{1}{2}$ lbs. of white wool, showing an increase of 40 per cent. in twelve years, which is certainly no mean improvement. The average annual clip of the Crossflat flock is now from $5\frac{1}{2}$ to 6 lbs., fully one-third more than blackfaced flocks on similar land generally average.

In 1872 Mr. Howatson purchased the estate of Glenbuck, a name which has since become famous amongst sheep-men all the world over. Although the lowest part of the land is 900 feet above sea-level, and the highest portion rises to an elevation of 1500 feet, it would be difficult to find such another sheep-farm more naturally adapted for the production of weight, quality, substance, and strength of constitution in the stock. Limestone is found near the surface on the low ground, and the grass in consequence possesses all the qualities necessary for the formation of bone and mutton ; but, like the greater part of the Ayrshire sheep-walks, Glenbuck has a good variety of pasture. The higher portions of the ground are covered with the finest quality of moss, which in good seasons comes away with the month of February, thus affording a bite to the sheep at the most critical time of the year, and rendering artificial feeding on the hill unnecessary, except in the case of a severe snowstorm. The rich low pasture makes bone, symmetry, and mutton ; but it has a tendency to produce a thick and rather short covering of wool ; while, on the other hand, mossy land favours the growth of a long wavy fleece, and kempy in the staple ; but a combination of the two varieties of grazing produces that symmetry in form and shape, the magnificent heads and faces, the clean mottled legs, and the stylish walk and appearance which the pure blackfaced sheep present to the eye when viewed on their native pastures. The fleece is both long in the staple and close at the roots ; the waviness peculiar to mossy pastures gives place to the proper kind of " pirl," and, to use the expression of a stapler, " it is all wool," and well adapted for protecting the animal at all seasons.

When the Cumberland Mining Company acquired the property of East Glenbuck, at the end of last century, it was fully stocked with blackfaced ewes, which were then, and for long after, the talk of the country-side for their great strength and substance, magnificent heads, and the extraordinary length and closeness of wool. The Cumberland Company failed after a few years spent in endeavouring to

develop the mineral resources of the estate; and when the
smash came the sheep stock were dispersed, and the seasons
being unpropitious at the time, the new tenant stocked the
place with wethers. The breaking-up of the ewe stock was,
however, an advantage to the neighbouring flockmasters,
some of the best ewe hirsels in the district at the present
day owing their superiority, it is said, to the old Glenbuck
breed. When the late Mr. James M'Kersie became tenant
of the grazings in 1811, he set about replacing the wether
with a ewe stock, and collected a few ewe lambs whose
descent could be traced back to the original stock; and, in
addition, he got some ewes and lambs from the flock at
Earlshaugh, near Moffat, which then, and for a long time
after, held the premier position among blackfaced sheep in
the south-western counties, the tups having gained all the
prizes at the first show the Highland Society held at Dumfries in 1830. Mr. M'Kersie started on the right track, and
being, in addition, an excellent judge of blackfaced sheep,
he was enabled to supplement his stock with the right blood
when occasion offered. It is only since coming into Mr.
Howatson's possession, however, that the Glenbuck flock
has achieved a national reputation.

With the Glenbuck flock Mr. Howatson has pursued
the same principles of breeding as he adopted so successfully at Crossflat. The farms of Glenbuck and Crossflat
are, indeed, within view of each other, and both being in
Mr. Howatson's own occupation, the two flocks are now
pretty nearly identical. He has been cautious, however, in
introducing new blood, preferring rather to mate different
tribes together, which, from the extent of the grazings and
the facilities afforded from having the two farms, he can
easily do. The only purchase made for the Glenbuck
flock was the ram "Seventy-one," so named because he cost
seventy-one guineas when a shearling. If any proof were
needed of Mr. Howatson's judgment in the buying of him,

it will be found in the fact that his son, "Seventy-two," took more prizes at the Highland Society's shows than any other blackfaced sheep. He was placed first at Inverness in 1883 as a shearling; first at the Centenary Show in 1884 as a two-year-old; first at Centenary Show as the best tup exhibited of any age; first in family group at Centenary Show; and first at Aberdeen in 1885 as aged ram.* Unfortunately he died from an accident in dipping during the summer of 1886. His g.-g.-g.-g.-sire was bred by Mr. M'Kersie at Glenbuck, and carried the first prize at the Sanquhar Show when ten years old; and it was this direct descent from the old breed which was probably the principal cause of the purchase. As at Crossflat, so at Glenbuck, the weight and quality of the wool-clip has materially increased with the improvement of the flock. The average weight of fleece at Glenbuck in 1880 was about 4 lbs.; in 1884 it had increased to $4\frac{3}{4}$ lbs.; in 1885 it was 5 lbs.; in 1886 it was $5\frac{1}{4}$ lbs. Mr. Howatson, it need scarcely be said, attaches great importance to his sheep carrying the sort of fleece best adapted for protecting the animal in a high, exposed, and stormy climate; and, as has been mentioned in another chapter, he has been instrumental in getting the Highland and Agricultural Society to offer prizes with the object of encouraging further improvement in that direction.

Having glanced at the means which Mr. Howatson has employed to improve the quality of his flock, we may proceed to notice a few points in the method of management which is pursued from year to year.

It is the usual practice to take five crops of lambs from hill ewes before they are drafted, but Mr. Howatson takes

* He was unquestionably the best blackfaced ram ever yet bred, and, as will be seen from the particulars which are given alongside our illustration of him on another page, he was the sire of many prize-winning sheep. 1.

only four crops from his before parting with them, for the threefold reason that the lambs are more vigorous and possessed of better constitutions when bred from young ewes than from old ones; that a flock can thereby be improved more rapidly; and that a better price is realised for the draft ewes when five than when kept until they are six years of age. The Glenbuck and Crossflat draft ewes are now sold at Lanark Auction Mart about the 6th of October.

After the ewes are drafted the whole flock is dipped, and the operation is repeated in January if the weather admits of it.

Every sheep in the flock has the letters " C·H " burned on the horn. The age-mark, the year of birth, ·5, ·6, ·7, denoting the year 85, or 86, or 87, &c., is also burned in the horn when the sheep is a year old.

It has been Mr. Howatson's practice to put the best lot of tups among the ewes in the first week of November— generally about the 6th of that month. This is about a week earlier than is customary in the district, but with his well-clad lambs the early lambing is not attended with any disadvantage. In three weeks or so the remainder of the rams are put out to the ewes, so as to pick up any that have not been served.

None of the ram lambs are castrated. The ewe lambs are sent back to the hill with their mothers, and are not weaned until about the middle of August. A few of the best young rams are retained for flock purposes, and the remainder, with any shearlings or older draft rams, are sold at the Glenbuck annual ram sale in the beginning of September, where they realise high prices. The old and antiquated system by which breeders concealed their light under a bushel has been entirely and absolutely discarded by Mr. Howatson, who has taken the public as much and as far into his confidence as it is possible for a breeder to do. The pedigree of every lamb offered is, to the fifth and sixth generation, detailed in the catalogue, and where this cannot be done the sire is noted as uncertain.

This system of selling the tups when lambs has much to recommend it. When rams are sold as shearlings, the exposer, in order to make his sheep look well, is often tempted to feed them so liberally that they become more fat than is desirable in sheep which are about to be put to service; hence the complaints one sometimes hears of the number of barren ewes which have been served by rams in this high condition. But it is impossible for the breeder to overfeed lambs when they are disposed of a few weeks after they are weaned, and the purchasers, getting possession of the tup lambs in their natural state of fatness, can afterwards feed them as they think proper.

Speaking of shearling and aged sheep of the blackfaced breed, unless we knew something about a ram we should be very careful not to buy it in September. At that season the fleece is too deceptive, and does not do good to the best quality, but improves the bad quality of skins. The best time for selling, or rather for buying, blackfaced sheep is in the month of May or June, their natural time for clipping. In September you cannot judge the quality of the wool; therefore, as a protection to buyers and improvement of the breed, the sales should be in May or June. Buyers, however, not breeders, are to blame for the September sales; for the latter cannot get the same prices in May. One reason for the September sales is, that it answers better for occupiers of unhealthy land to take their tups home at that season. But surely it would be better to purchase the rams in May with their full and natural fleeces; and if the ground they are going to is unhealthy, make an arrangement with the seller to keep them until the autumn; and then, if they are in high condition, there is sufficient time to reduce them, and prepare them for the tupping season in November; or, purchase well-

G

grown tup lambs in September, winter them as may be thought best in the low country, and prepare them for the following season's use, according to the tastes and wishes of the proprietor. This, however, is partly digressive.

The ewe lambs at Glenbuck are also weaned in August, and those not intended to be retained to keep up the stock are disposed of at the annual ram sale in September, or are sold to farmers who winter them on the pastures which have been grazed by dairy cows during the summer; and in the following spring they are sold by them to be used for breeding from, many of them going to Galloway, the north of England, and elsewhere. The stock ewe lambs are dipped and then sent back to the hill until the 12th of October, when they go away to the low country to be wintered on grass, returning on the 1st of April. The cost of this wintering is 7s. to 8s. per head.

The ewe hoggs are shorn in the second week of June, the utmost care being taken to mark any of them which in any respect, but especially from their coats, are objectionable as breeders, so that they may be drafted from the stock and sold towards the end of the year. The ewes are clipped in the latter part of July. This would be considered late in some localities, but Mr. Howatson considers it is of advantage to be late in shearing the ewes, provided they are so well bred as to carry their wool until the new wool is well raised. Early clipping drives the milk from the ewes, as well as reduces them in condition, if the immediately succeeding weather is either hot or cold. The only drawback to this is, that in very warm weather the unshorn sheep have a disposition to roll, and sometimes fall "awlt," which necessitates greater watchfulness on the part of the shepherd.

· For experimental purposes the following blackfaced rams were last summer bought by Mr. Howatson, and are being used at Glenbuck this season, viz. :—The first prize shearling ram at the Royal Newcastle Show, bred by Professor M'Cracken, price £50; the second prize shearling ram at West Linton Show, bred by Mr. Woddrop of Garvald, price £50; the highest-priced ram at the Edinburgh ram sales, bred by His Grace the Duke of Argyll, price £52, 10s.; a shearling ram selected from Mr. Murray's flock at Parkhall, price £75; a three-shear ram, "The Bailie," bred by Mr. Foyer, Knowehead, Campsie (used as a stud ram at Benmore), bought at Perth sales for £26, 10s. Besides the above rams ten ewes, selected from the whole of the stock at Knowehead, have also gone to Glenbuck. It is said that Mr. M'Indoe, Knowehead, has secured the highest price ever paid for ten ewes by giving this selection.

Since getting possession of Glenbuck, Mr. Howatson has drained and limed about 150 acres of the hill land; and, with the exception of one small spot where there is a surface mossy heath resting on a cold clay, the improvement and sweetening of the pasture which has followed the application of the lime is truly wonderful; the black heath having on such places for the most part entirely disappeared, and given place to a healthy green sward of natural grasses and clovers. There being a want of natural shelter on the hill, Mr. Howatson planted a strip of plantation on one of the most exposed parts; but while the young trees are growing fairly well at the lower end, they seem to have died off entirely at about half-way up the hill. Several old stells on the hill which had been allowed to go to ruins have been rebuilt, in circular form, where the different cuts of sheep can be foddered with hay in case of a heavy snowstorm. But Mr. Howatson has not been content with merely improving the flock and its pasture. The coal-mines on the property have been developed, and for some years have shown a large annual output. The old farm-house, too, has given place to a splendid mansion, while on the brae-face behind are gardens and vineries, which produce grapes and peaches of the finest flavour and quality.

On the occasion of leaving Daldorch to take up his residence at the new mansion of Glenbuck, in July 1880, Mr. Howatson was entertained at a public dinner in the Crown Hotel, Catrine, and presented with oil paintings of himself and Mrs. Howatson. Both portraits (now occupying prominent positions in the dining-room at Glenbuck) are excellent likenesses, as they undoubtedly are fine works of art. They were painted by Mr. Johnstone of Manchester, a native of Auchinleck, Mr. Howatson's native parish. On a plate attached to Mr. Howatson's portrait is the following inscription :—

" PRESENTED TO CHARLES HOWATSON, Esq., of Dornel and Glenbuck, by friends in Catrine and the surrounding district, in token of esteem, and in appreciation of his public services while residing at Daldorch House."

On a similar plate attached to Mrs. Howatson's portrait is a corresponding inscription testifying to the "grateful remembrance of her many acts of kindness to the people of the district."

It will not be out of place here to mention that, on the 21st of March 1884, Mr. and Mrs. Howatson celebrated their SILVER WEDDING, on which occasion they were presented with a silver casket containing an illuminated address by the inhabitants of Glenbuck and Muirkirk. In acknowledging the presentation Mr. Howatson said :—" You have spoken of me in terms far too flattering. You have also referred to Mrs. Howatson in glowing terms, and I can assure you that, however it may be in my case, you have in no way over-estimated her virtues. Her kindly disposition, forbearing nature, and sweet temper have made for me a happy home ; and it is owing to this that I am able to say to you to-night, that for the twenty-five years I have been a married man I have never had an angry word in my house ; and which makes the inscription on your programme a reality, for ' Happy we've been aye thegither.' "

[The foregoing pages were in type some months before the lamented death of Mrs. Howatson in October 1887, after more than a year's severe illness, borne with exemplary patience and fortitude. In her the poor of Muirkirk and district have indeed lost a friend.]

CHAPTER XIV.

SYSTEMS OF MANAGEMENT.

HILL sheep-farming admits of few different systems of management. The pursuit is virtually a grazing and breeding rather than a fattening occupation. Even then the systems of grazing are limited as regards the different kinds of stock which can be profitably kept. Practically, the business is reduced to two distinct branches—(1) a ewe flock, and (2) a wedder flock. The distinguishing feature between these systems is regulated, not so much by which of the two shall be most profitable, as by which of them is best adapted for the situation. But, of course, the system best adapted for the situation is also certain to be the most profitable. Breeding flocks are best adapted to the lower and better quality of farms, while the higher grounds are more suitable for wedder sheep, as may be readily inferred when it is stated that at the period of lambing the highest grazings are frequently covered with snow—the new growth of grass, so essential for ewes at that season of the year, having not then appeared. On the other hand, good strong wedders take less hurt from the rough weather of winter, and are better able to rustle for a livelihood in deep snows. In late untoward seasons the losses among young lambs are sometimes extremely heavy, and on that account it is generally better policy to keep wedders where the danger to lambs becomes extreme. Ewe flocks are undoubtedly the most profitable in good seasons, even on high ground; but the risk of breeding has to be considered, and, as a rule, it is found more profitable to adhere to one system or the other, according to the altitude and exposure of the farm.

A Ewe Flock.—We cannot better describe the routine of management on a blackfaced sheep-farm than by describing a farm of some 9000 acres with which we are well acquainted. It is calculated to carry about 3700 breeding ewes, with or about a hundred head of cattle during summer. Four shepherds are employed on this farm, living in houses situated at convenient parts of the farm for attending to their several hirsels. Each hirsel of sheep is composed of all the various ages necessary for their own maintenance in about the following proportionate numbers :—

> 800 hoggs.
> 760 gimmers.
> 740 three-year-old ewes.
> 720 four-year-old   „
> 700 five-year-old   „
>   60 tups.
> ———
> 3780

Tups.—The tups are grazed on enclosed land near to the farmhouse. There are no packs on the farm, the shepherds being paid a money wage. There are really more sheep on the farm than stated, as the object is to keep these numbers up after deducting losses by death and other casualties. Lambing commences about 20th April, and during that season two extra shepherds are employed, each of whom takes a share of the ground herded by two of the regular shepherds. In about a month the lambing is nearly over. As a rule, the best of the gimmers only are allowed to bring lambs, and in some years when they are small generally none of them are put to the tup. To an outsider this would seem a foolish policy, but as a gimmer lamb is generally weakly and of little value, it is found more profitable to let her go without one, as the crop of wool will be greater and a good constitution established for after years. Small blackfaced lambs bring only a small price, and it is only on the better class of hill farms where it is prudent to breed them at the risk of spoiling them for after years.

Lambing.—Lambing-time begins about the middle of April, and the grass on the hills at that period is only beginning to sprout. In late seasons no growth of grass is visible even then, and when such is the case the outlook for the young lambs is dismal in the extreme, especially when the ground happens to be still white with snow. It may be imagined that under such conditions tender weakly lambs stand a poor chance of surviving; yet when they are well bred, which also means that they are well covered with wool when they are born, it is surprising how much cold and privation they can endure. In good seasons blackfaced ewes rear on an average a crop of lambs equal to 95 per cent., but in bad years, with stormy weather at lambing time, the lambs raised may not average more than 75 or 80 per cent. of the ewes. The condition of the flock, and the prevailing weather, are the ruling elements at this season on a hill farm. The ewes lamb upon the open hill, and require little or no assistance. The young lambs soon rise and follow their mothers, and when the weather is fine the lambing is a comparatively easy operation. It is only during cold and barren seasons that difficulty arises. Then it is, too, that the advantage of stells and kebhouses becomes apparent. The loss of lambs is sometimes enormous, not from any fault of the sheep, but simply owing to the wild tempests of sleet and snow to which they are exposed. Before lambing commences, the thinnest ewes in the flock are picked out by the shepherds and taken into the parks and meadows, and thus a great deal of labour and trouble is saved, for

they are then more easily looked after. When in good condition blackfaced ewes are kind mothers, but when thin
their maternal instincts are less intense, and, like all other
breeds, they are apt to drop their lambs and pay no attention to them afterwards. This is one of the greatest difficulties the shepherds meet with. "You can take a horse
to water, but that is not to say you can make him drink."
The same holds good with the ewe. Once she leaves her
lamb it is sometimes very difficult to induce her to again
acknowledge it.

Tup Lambs.—When about a month old the tup lambs
are castrated, a few of the best being reserved for sale or
breeding purposes. Tup lambs are allowed to follow their
dams upon the hill till weaning, after which they are better
treated than the ordinary stock, but no attempt at over-feeding is practised.

Clipping.—Some weeks previous to shearing, the shepherds have an anxious time, as it is then that the ewes are
apt to get "awalt." When the grass is very succulent
and a ewe happens to roll on her back, death often ensues
very quickly. Nothing vexes a shepherd so much as death
from this cause. Clipping takes place about the first week
in July. Some farmers shear the hoggs and eild sheep a
fortnight or three weeks earlier than the ewes with lambs,
as they consider it injurious to clip milch ewes too soon in
the season. Blackfaced sheep are not washed previous to
clipping, unless the facilities for doing so are very great. In
many parts of Scotland it is a difficult task to gather the
sheep from the hills for such an operation; and some wool-
brokers tell the farmers that blackfaced wool is not improved
in value by washing; so that in many cases—perhaps two-
thirds—the operation is dispensed with. The shepherds of
adjoining farms assist each other in this work. The weight

of the clip varies very much according to the season. From three to five pounds of wool per sheep, according to quality of land, would be near the mark. On the farm to which we refer the ewes will average about four pounds each. A good clipper can get over fifty to seventy sheep a day, and as the men work with a will the flock are soon minus their coats. Like most mountain sheep, blackfaced are shorn across the rib, not with it. As a rule, the work is coarsely done; all hill shepherds like to leave a short growth of wool to protect the sheep from sudden storms of cold wet nights. As the fleeces are taken off, they are properly tagged of all soiled wool, neatly rolled up, and stored in a barn or wool-house, ready for inspection by the wool buyers, who periodically come and see the "clip." Then comes a short time of rest for the shepherd, a period for *otium cum dignitate*. A good deal of wool is still sold by "character" at old-established markets in the north of Scotland; and there are many farmers in the Highlands who have never sold their wool in any other way, nor have they any wish to change the practice.

Weaning.—About the first day of August, when the lambs are taken from their mothers. From 2160 ewes about 1950 lambs are weaned, a few more in good seasons, and in bad ones a good many fewer. The system followed is to draw out 800 of the best ewe lambs, which are put on one of the hirsels, or part of one, which has been cleared of stock previously, and the grass is clean and nice. Sometimes this lot is sent away for three weeks or a month to another part of the country for the sake of change, as well as to let the land on which they are to be wintered get a fair start. Others again wean their ewe lambs quite early, and put them on a pasture apart from their mothers— usually the best the farm provides—until the time for sending them to winter quarters arrives. A large portion of the hill is thus unstocked during winter. It is fresh and clean for the hoggs and dinmonts in the spring when they return to it. The cost of keeping them thus is about 3½d. per week, so that a heavy bill comes against the farm for their winter keep. On farms where the hoggs can be wintered at home, the lambs are merely herded apart from the ewes till such time as they are fairly weaned, when they are put back to their old ground, and allowed to graze along with their dams.

Ewe Lambs.—The management of the ewe lambs is the most delicate portion of the hill-farmer's work. After being weaned for a fortnight they are allowed to go back to their mothers, but it is only on sound, good places that this practice can be followed. Where there is a mixture of "white" ground and heath they generally do well, but as there is a great want of the latter, the ewe-hoggs are often in bad condition during spring, and the death-rate is very large. The first point is clean good pasture, and during the autumn months they should be moved about, regularly driven on to the low ground through the day, and sent back on to the hill-tops at night, whether they are herded separately or along with the old sheep. They thus get a change of food, which is beneficial to all classes of stock. Sickness is often prevalent up to new-year, then during the spring months they often begin to pine and waste away. The only cure for this disease is a change; there is the great difficulty, and one that is not easily overcome. The place to which we refer rests at one part upon a granite and at another upon a sandstone formation. A change from one place to another in this case makes a great difference, and often puts the hoggs in capital order; but the most effectual way, when it can be carried out, is to bring them down to the low lands, and put them on artificial grass. A very sanguine shepherd once mentioned to us that the very smell of young grass cured them; certain it is that in a few days they pick up wonderfully, and in a fortnight's time it is either kill or cure—nearly always the latter. If the month of March is once got over, there is little danger after that time. Another curious fact connected with "hogging" ewe lambs is that they cannot consecutively be wintered upon one place. They must be put on fresh ground every year. Whether it be that they eat certain grasses peculiar to them so heartily that there is not enough for them the following year, we cannot positively say, but it is a very likely explanation. The death-rate amongst ewe lambs is often very large, and on an average will not be less than 10 per cent.; but, once past this stage, there is little loss in after years—not above two or three per cent. yearly. A winter's keep in the lowlands, the hoggs getting nothing but grass, and hay in case of snow, costs from 7s. to 8s. a head, which is a heavy tax on the hill farmer. When prices for sheep and wool were high he could better afford it, but for some years bygone the cost of wintering has been ruination to those having to pay for it. How to overcome this difficulty is at present the hill farmer's most earnest study. Something will have to be done if rents are not permanently lowered. Keeping a lighter stock is no remedy. The best plan, perhaps, would be to make a full supply of ensilage and hay and keep the hoggs at home, bringing them in from the higher ground; and even then the stock would have to be reduced in number, but nothing else seems practicable under the circumstances.

Hay for Winter.—During the month of August the hay harvest commences. Of late years great attention has been paid to the making of hay. It is sometimes cut out of the bogs on the open hill, but more generally from enclosed meadows and fields specially made for the purpose. Good hay is of enormous value to the stock farmers. It is his only hope in deep snow, when the pasture is "sealed" by intense frost. Many a flock has been kept in good condition through a plentiful supply of hay. It is, of course, seldom that sheep cannot subsist by some means or other, but we have seen occasions—certainly few and far between —when hay was absolutely necessary to keep in life. It is always safe policy in stormy weather to supplement the natural food with hay. Blackfaces being naturally very hardy, they require less artificial feeding in winter than almost any other breed of mountain sheep; yet in excessively severe winters the prudent manager does not leave his sheep to forage for themselves until it is too late to help them. So long as the snow does not get too deep, or is not frozen hard, they take little harm. Blackfaced sheep are excellent workers in the snow, and will toil bravely for a sustenance under the most trying circumstances. Hand-feeding is only resorted to when it cannot be longer avoided; and in that case the sheep are either removed to a lower district or fed on hay at home. Hay is still considered the best of winter foods for hill sheep, but silage is likely to take its place very largely in the damp regions of the Western Highlands, where hay-making is frequently seriously interrupted by wet or misty weather. Corn is only fed to blackfaced sheep in winter when hay would be ineffectual in maintaining ewes in a condition fit to rear their lambs. Those so treated, however, are usually drafted the following autumn. An open season is more effective than any system of artificial feeding in bringing these hill sheep through the winter, and if given half a chance they seldom need or ask anything else.

Autumn Sales.—The month of October is the most important of the year in a financial point of view. The wool is generally sold, delivered, and paid for in August or September; but during the above month the stock farmer disposes of his autumn casts—that is, his draft ewes and his wethers. Thus the 700 five-year-old ewes are disposed of. This class of stock is extremely valuable. They are invariably kept on for another year, principally by farmers in the south, who cross them with Leicester rams, producing lambs for the fat market; in fact, on good pasture they feed both themselves and their progeny fat. For crossing purposes and rearing fat lambs, blackfaced ewes stand unrivalled; but it is only very recently that English buyers of Scotch sheep have found this out. The change from the high mountains of Scotland to rich English pasture agrees so well with the blackfaces that they fatten without the aid of extraneous foods of any kind, and the great weights to which their lambs grow is quite surprising. We cannot too strongly advise English sheep feeders, who go to Scotland for lambs and ewes in the autumn, if they have not done so before, to give the blackfaced breed a trial, for they will never regret doing so. The wether lambs, together with the second and inferior ewe lambs, are sold direct from their dams in September. The "shott" lambs are disposed of generally at a very small price. The aged ewes are drafted and sold about a month later than the lambs. As a rule, the ewes are kept till six years old, but not a few breeders dispose of them a year younger. Thus each ewe rears four or five lambs, after which she is drafted.

Putting out the Rams.—The winter's work fairly begun, the most important operation is setting the rams to the hill. This is done about the 20th November, and each ram is supposed to serve fifty ewes. Rams on those hills need to be in their hunting condition, able to follow the ewes on daily journey from mountain top to sheltered vales and rich grassy holms, and back to their airy resting-places when night throws her mantle o'er the scene. The object of all flockmasters is to use young tups, as they are surer and more active, and many prefer lamb rams to shearlings; only a few of the best are kept for another year, the remainder being fed. Careful breeders select the best of their rams to turn out first, and after these have been three weeks with the ewes they are brought in, and the second-rate rams, which number about one to three of the first lot, are turned out to pick up any ewes that may have been missed by the others. This plan, while being more certain of getting all the ewes in lamb, saves the best rams from being overworked or getting too much reduced in condition. The rams are finally withdrawn at the end of the year.

Dipping.—The last important work of the year is dipping. All lambs are dipped at weaning, but ewes are dipped either in autumn or early spring.

A Wedder Flock.—Wedder flocks are much more easily managed than where breeding is carried on. They are delegated to the highest mountain ranges, and in winter considerable watchfulness is necessary on the part of the owner to see that the sheep are not starved beyond what they are well able to bear. A regular wedder flock of 3000 sheep is composed of either two, three, or four ages, thus:—

$$
\begin{array}{ll}
\left.\begin{array}{l}\text{1500 hoggs} \\ \text{1500 two-year-olds}\end{array}\right\} & \text{two ages.} \\[1em]
\left.\begin{array}{l}\text{1000 hoggs} \\ \text{1000 two-year-olds} \\ \text{1000 three-year-olds}\end{array}\right\} & \text{three ages.} \\[1.5em]
\left.\begin{array}{l}\text{750 hoggs} \\ \text{750 two-year-olds} \\ \text{750 three-year-olds} \\ \text{750 four-year-olds}\end{array}\right\} & \text{four ages.}
\end{array}
$$

Which of these systems it may be most profitable to follow is determined a good deal by the character of the grazing. Where the farm is high-lying or of a poor quality for grazing, and so exposed that the hoggs cannot be maintained at home in winter, then it will be seen that 750 hoggs are very much more easily wintered away than 1500. The farm may indeed be unsuitable for carrying sheep of any kind during winter, or it may be that only strong three or four year-old wethers can be kept on it. Wintering away is expensive, and in deciding which of the ages to keep it all depends whether the outlay is justified by the receipts. The poorer the farm is in respect to the quality of sheep it produces, and the more difficult it is to winter them at home, then the older-age system is likely to pay best. On the other hand, the better quality of sheep produced and the more favourable it is for wintering, the younger ages will generally be most profitable.

There is, however, not the same necessity for keeping blackfaced wedders till three and four years old as once was. The breed has been much improved as regards the age at which they can be profitably fattened. At one time buyers of hill wedders would scarcely look at a two-year-old sheep for fattening purposes, and as a consequence graziers were obliged to keep them until they had reached a maturer age. But now a two-year-old wedder is worth quite as much as one a year older, providing the weight and size of both are about equal. The younger sheep should, in fact, be worth a trifle more, because it has still the advantage of growth in its favour, while the other has nearly reached to full size. Every intelligent feeder is now aware that a young growing animal gives a better return for food consumed than one that has attained maturity. The latter doubtless fattens more rapidly; but then the young animal, in fattening, at the same time also increases in size, and the combined progress it makes in size and fatness gives a greater weight on the aggregate than can be yielded in fat only without any additional growth. For this reason young sheep suitable for feeding purposes find a readier market than formerly, and breeders are doing all they can to improve their flocks in early maturity.

Wedder Flocks are kept up by buying lambs in the autumn, the numbers required being subject to the age at which they are drafted. The hoggs on such farms are usually to winter away, the home ground being too stormy for them in winter. Wintering is found on cattle or dairy farms in the Lowlands. Grass only, with hay during a snowstorm, is the usual winter fare for hill hoggs; but some stipulate with the arable farmer who supplies the grass to provide also a certain quantity of turnips. This makes the hoggs stronger in the bone, and gives them a bigger and better appearance when they come to be sold; but it is doubtful if so expensive wintering always pays. However, that is a point which can best be decided by those concerned. It may pay, or it may not, according to the benefits derived. The better the hoggs are treated in winter, the better must they also be done by in summer. It never pays to turnip wedder hoggs in winter to be afterwards grazed on inferior pasture.

A Mixed Flock.—On some farms a mixed stock of ewes and wedders is often more profitable than confining the system to one or the other only. The wedders are kept on the highest part of the farm and the ewes on the lower. The wedder lambs which are bred from the ewe hirsels, instead of being sold at weaning, are drafted on to the wedder hirsels, where they are grazed until two or three years of age. This system prevails very largely throughout the country. The lambs, being bred on the farm, are more healthy and generally live better than when they are bought in, especially if the land is at all subject to disease. Another advantage of keeping a wedder hirsel in conjunction with a ewe flock is found in the better average quality of

H

the lambs produced.   They are few in number, but being reared on the best parts of the farm only, they are not reduced in average quality by those from the higher and inferior grounds, which are often better adapted for growing a good old sheep than a good young one.   The management of a mixed flock does not vary from either of the two distinct systems of a ewe or wedder flock already described. It is therefore unnecessary to repeat the routine of work for the year, as it may be fully gathered from the previous remarks.

CHAPTER XV.

LAMBING.

THE lambing season is the most important event of the whole year on a hill farm. It usually commences about the 20th of April, and finishes about the end of May. Months before it actually occurs the shepherds are deeply interested in the coming ordeal, and the weather is their chief theme of conversation. In severe winters and late, cold springs, hill ewes are sometimes sorely reduced in condition, which unfits them for rearing their lambs; and in these circumstances the lambing is usually a period of much anxiety to the flock-masters, as well as one of great trouble to the shepherds. On the other hand, in good seasons little difficulty or loss is experienced, and the event is more a pleasure than a season of anxiety to all concerned.

The advantages of bringing the sheep through winter in good order is at no time more apparent than when the lambing begins. At that season, whatever the weather may be, condition in the flock speaks plain and forcibly.

In other words, when the ewes are strong and full-fleshed they are able to satisfy their offspring with plenty of milk, which gives them strength and fitness to endure the cold blasts to which they are exposed. But should the ewes be lean and " poverty-struck," they are hardly able to fend for themselves, much less to give nourishment to their lambs, which become a source of incessant care to maintain alive by hand-nursing, many of them succumbing at an early age, or eventually growing up miserable and worthless creatures.

Of all breeds, blackfaces are pre-eminently the most hardy, and cause least trouble at lambing. And not only do they cause least trouble, but they will rear twenty per cent. more lambs than their next neighbours the cheviots, having the very same opportunities. It is for this reason, more than any other, perhaps, that they have ousted the cheviots from so many of their grazings. A blackfaced ewe can maintain her condition where a cheviot would starve; and at lambing-time she not only gives more milk, but her lamb possesses the same qualities of endurance, which enables it to withstand the sleety blast and survive where a cheviot would perish. The activity of blackfaced lambs is truly wonderful. While those of other breeds, after being dropped, require from fifteen to thirty minutes and longer to get upon their feet, the blackfaces are up and sucking in five; and this little qualification is of great moment as a means of preserving their life when born in the midst of snow or rain. Preparation for lambing commences some months before the date on which it is expected to arrive.

This consists of an examination of the whole flock, when any ewes that are considered too weak or thin to lamb successfully on the hill are drawn out and taken into the enclosures near the house, where they are fed on hay and turnips, so as to fit them for rearing their lambs. In certain seasons the majority of the ewes would be all the better of this treatment; but when poverty afflicts the whole flock, the results will be disastrous in whatever way they are managed; and it is sometimes difficult to say whether extra feeding should be resorted to or not. Any move will pay better than allowing the sheep to die of starvation; but, speaking broadly, the worst cases only should be hand-fed. Any ewe apparently strong enough to bring her lamb without extra feeding had best remain on the hill and take her

chance. To the inexperienced this may seem strange coun-
sel, but with hill sheep more than present wants have to be
taken into consideration. Hill pasture is insufficiently rich
to afterwards maintain artificially fed sheep in a thriving
condition; and when once such food has been given, it has
to be continued every year. For a short time in summer
the hand-fed sheep will thrive on the hill all right, but as
soon as winter sets in they fall away in condition and look
for better keep. Hay and turnips are less injurious in their
after effects than cake or corn; and the latter should only
be given in cases of extreme debility, when the former would
be ineffectual in bringing the sheep up to the required state
of health and vigour. The shepherd requires to be careful
to disturb his sheep as little as possible at this time, only
guiding them to those portions of ground where the most
suitable food can be found. Before lambing commences it
is also necessary to have all the keb-houses and shelters put
in good repair.

Every little convenience is of importance, and the farmer
ought to make a point of providing the necessary materials,
which do not cost much, such as a few boards and hammer
and nails. A careful shepherd will not fail to make good
use of them; and he will also make clean and comfortable
all the pens in which he may afterwards have to confine any
of his ewes. Small troughs for feeding sick or penned-up
sheep should also be provided, as well as the necessary food.
It is no secret that some shepherds think nothing of shutting
up a ewe in out-of-the-way keb-houses for days together
without giving her a bite to eat, simply because he has
nothing provided for the purpose. Such treatment is not
only hurtful to the ewe, but it defeats the very object for
which she is generally confined, viz., that of adopting her
own or a strange lamb. The way to make a ewe take
quickly to a lamb is to give her plenty to eat, to bring as
much milk into her udder as possible. She will then turn

to the lamb for relief; but when she is starved of food the process of adoption is slow, and more liable to result in failure altogether. These are, perhaps, little points, but they are of more consequence than some may think.

When lambing begins the shepherd requires to see his flock three times a day. His first round is made at early dawn, before the sheep have left their " moorings," when any requiring attention can be readily noticed. Some shepherds make this trip before breakfast, but this is not a good plan to adopt. When a shepherd leaves his house he never knows how long he may be detained; and going out hungry may cause him to leave his work when he ought not to do so, especially in bad weather. On returning from his rounds he brings home any ewe that has lost her lamb. Having keb-houses at various parts of the hill is of immense advantage at this time, and saves not only the shepherd a lot of unnecessary work, but is much better for the sheep every way. Then there will be a number of such stock in the hospital individually requiring careful treatment, all of which he needs to see before returning to the house for a meal. There is no time for rest during the day, and no sooner is one journey finished than he starts on another, repeating the same morning, noon, and evening. Much depends on the weather, and the worse it is the more need there is for exertion and daily perseverance, which the shepherds, as a rule, never grudge in behalf of their flocks. In order to induce a ewe to take a stranger lamb under her charge, the skin of her own dead lamb is flayed and put on another lamb, when the smell of the old skin is usually enough to deceive and induce her to take kindly to the newcomer. Instead of adopting this method, which involves more or less labour, sometimes the ewe is milked, and the milk is rubbed over the skin of the lamb that is to be transferred to her care; and it is found that the smell of her own milk has the same deceptive effect as the smell of the old skin. The different natures and dispositions of sheep are seen in their treatment of their young perhaps more clearly than in anything else. Like their rational superiors, some are full of natural affection, while others have little of the milk of kindness about them. Some, in defence of their young, will boldly defy a shepherd dog, from which they would, at other times, have fled in terror; while others are careless about their offspring, and when they lose a lamb will comport themselves with as much indifference as if nothing had happened. When any of those sheep which belong to what we may call the affectionate type have lost a lamb by death, it requires not a little tact and perseverance to manage them in a proper manner. They will stand over their dead offspring, cry piteously, and although no response is given to their plaintive bleatings, no force or persuasion will induce them to leave the spot. In a case of this kind the shepherd occasionally has recourse to the expedient of tying a long string round the dead lamb, and by pulling it gently along the grass, the inconsolable mother, thinking it to be alive, follows it until it is secure within the fold.

The disorders most commonly met with at lambing among blackfaced ewes are awkward presentations, inflammation, and subsequent protrusion of the womb. The first may be due to natural causes, but more often it is the result of accident. The ewes are liable to get injured, or an abrupt turn with a dog may result in wrong presentation, if not premature delivery of the fœtus. All probable mishaps require to be guarded against, and prevented as far as possible. Inflammation of the uterus and other affections of the womb are due to such causes as cold, injury during parturition, and contagious influences.

Ewes that have had difficult delivery are, of course, most liable to septic attacks; and it is very essential to protect all such cases from undue exposure by placing them under shelter. Injury during parturition may be largely prevented by a skilful attendant. The first essential in all operations of the kind is clean hands, which, after washing, should be rubbed over with pure oil. The mucous membrane of the uterine passage is in an excited condition at this time, and prone to suffer from contact with infectious matter of any kind. Absorption of septic germs takes place very rapidly, and blood-poisoning is the natural consequence. The horns of the lambs frequently cause serious injury to the ewe in lambing, and the shepherd should provide himself with a wire or instrument with which he can render more effective assistance in delivery than he could with his hands only.

The remedy for inflammation, or septic infection of the uterus, as it is called, is a dilution of carbolic acid. According to the severity of the case, mix from five to twenty parts of Gallipoli oil in one part of Calvert's best carbolic, and pour about two tablespoonfuls of the mixture into the womb. It is also important to dress all the outer parts which appear inflamed. This operation is repeated every third or fourth hour, as may be required. With so simple, cheap, and efficacious a remedy at hand no ewe need be lost at lambing. Its successful use, however, depends very much on early application. Any suspicious case should be treated with a weak solution immediately after lambing.

It frequently happens that many of the ewes are unable to satisfy their offspring with a sufficiency of milk the first few days after lambing. This casualty may even occur to ewes in otherwise good health and condition, as well as to those that are poor and thin-fleshed. The remedy usually adopted is to feed the lambs on cows' milk until such time as the ewe recovers her natural functions. It is a curious fact, however, that the milk of some cows is decidedly injurious to young lambs. We have often noticed hand-nursed lambs to thrive very indifferently, and in some cases, where their chief nutriment was from the bottle, to pine away and die, sometimes a considerable loss resulting without the cause being detected.

But feeding cows' milk also affects the health of the lambs at a later stage of their life. On farms that are naturally subject to disease it is an axiom with the shepherds "never to feed cows' milk." They are well aware it is only time and trouble wasted. The lamb's life may be saved in the first instance, but within the course of the next four months it is certain to take some fatal disorder.

A remarkable instance of this kind was once related to us by an observant shepherd. A very stormy day occurred during the lambing season, and to save as many lives as possible he carried upwards of forty newly dropped lambs into a house for shelter; for want of room the ewes were left outside. The storm continued longer than was anticipated, and to have turned the lambs out in it, hungry as they were, would have been doubly fatal. He procured cow's milk and administered it to the lambs, which were retained in the house until the following day. After some difficulty in getting all the ewes to adopt their lambs, he was highly gratified to think he had successfully saved so many lives. " But," said he, "would you believe it? not one of these lambs lived to be weaned."

While it is true that excellent lambs may be reared solely upon cows' milk, it is also admitted that when fed as a supplementary ration to that which the ewe supplies its effects are more or less injurious. That is to say, the cows' milk does not agree with the lamb so well as the natural food, and should only be given in cases of dire necessity.

There is a vast difference, however, in the effect produced by the milk of various cows; and the period of calving appears to have a great deal to do in determining the result. Some folks prefer a fresh-calved cow for use at lambing-time, whilst others select the oldest they can find, choosing a farrow one if possible. But it is impossible to

say whether the milk from new or old calved cow is most
suitable for young lambs. That will greatly depend on
how the cow is being fed; a medium age is perhaps the
safest to adopt. When more than ordinary misfortune is
apparent amongst hand-nursed lambs, a change of milk
may have a beneficial effect.

The ailments of hill lambs as a rule, however, are few and
seldom. Their first and most fatal enemy is cold or hunger.
For reviving chilled lambs the shepherd carries constantly
in his bosom a bottle of warm milk, and sometimes another
containing gin or whisky, of which he supplies a mouthful
in extreme cases of weakness. Lambs that are really pros-
trate with cold have to be carried to some place of shelter.
Very often the shepherd's kitchen is turned into an hospital
for subjects of this kind. In a stormy day it is not unusual
to see twenty or thirty shivering lambs by his fireside, which
his wife or children attend to while he is away on his rounds.
Lambs are, however, never taken from their dams if it can
possibly be avoided. There is often some difficulty in getting the ewes to own them again, the natural odour by
which they are recognised by the mother having been dissi-
pated by the heat of the fire, or from coming in contact
with others of a different smell. A better method of reviving chilled lambs than warming them by the fire is to dip
them in a tub of warm water, then, after wiping dry, wrap in
a woollen cloth, and leave them beside the ewes in the kebhouse. On recovery, care must be taken to accustom the
lambs gradually to out-door life. A sunny noon is a favourable time to set them out, but if the weather continues cold
they should be housed for a few nights, until they are
strong enough to withstand the elements to which they are
exposed.

The lambing season is a time of great excitement among
the shepherds. Some, indeed, become so anxious and nervous
that they cannot be said to enjoy a single night's refreshing

sleep during the whole time. This is more especially the case with those shepherds who reside on led farms, and on whom devolves all responsibility connected with the sheep. Yet their anxiety is more the result of their own temperament and a sense of their position than of their knowledge regarding any fault-finding tendency or exacting demands of their masters. Their masters, for the most part, place entire confidence in them, as is abundantly evidenced from the fact that they seldom pay a visit to their farms during the whole year. And this confidence is well placed; for it is found, on the most careful scrutiny, that the sheep are as well cared for and all other things as well managed as if they were themselves always on the spot. After a month the flock will have mostly lambed, with the exception of those which missed the first tupping. The shepherd then collects all that have not lambed, and after setting the eild ones free, he brings the late lambers into the meadows or enclosures, where he can attend to them more easily than on the hill. By this time, if the weather is mild, the hill pasture should be well sprung, and the leanest of the flock, brought in previously, may be again turned out.

The ewes with twins, which will not be many, at the end of lambing-time are kept in the longest of any; but they too have to be cleared out as soon as possible, to prepare the enclosed land for cutting a crop of hay.

Lambing is then over. The assistants, if any, are disengaged, and the shepherd resumes the entire charge of his hirsel.

His duties are not much lighter from now till he gets his sheep clipped. In the month of June the sheep are apt to "awald," that is, to roll on account of "keds," and, being rough in their wool, "lie awkward," or square on their backs, from which position they are unable to rise, and if not rescued they very soon die.

Duration of Pregnancy.—The period of gestation in the ewe is generally estimated at twenty-one weeks, but the duration of pregnancy varies as follows :—

4½ months, or 135 days,	.	.	.	Premature labour.
4⅘ „ „ 144 „	.	.	.	Regular „
5⅓ „ „ 160 „	.	.	.	Protracted „

LAMBING TABLE.

[Showing at one view when twenty-one days will expire from the 1st and 14th of any month.]

From		To		From		To	
January	1.	May	27.	July	1.	November	25.
„	14.	June	10.	„	14.	December	9.
February	1.	„	28.	August	1.	„	26.
„	14.	July	12.	„	14.	January	8.
March	1.	„	26.	September	1.	„	26.
„	14.	August	8.	„	14.	February	9.
April	1.	„	26.	October	1.	„	25.
„	14.	September	8.	„	14.	March	10.
May	1.	„	22.	November	1.	„	26.
„	14.	October	8.	„	14.	April	9.
June	1.	„	25.	December	1.	„	25.
„	14.	November	8.	„	14.	May	9.

Emasculation.—As soon as lambing is finished, the whole flock are gathered into the fold for the purpose of castrating the tup lambs. A good day in settled weather is chosen for this operation—one neither very warm nor yet too cold. In gathering the sheep, the shepherd drives them very slowly, and carefully avoids overheating the lambs. Care is also taken not to crush the lambs which are as yet young and tender, plenty of room being allowed in the pens. Most people prefer to "cut" lambs when about a month old, but in hill flocks only one gathering is considered good for the sheep, and on that account the operation is performed on all ages from one to six weeks

old. It seldom happens that any great casualty results from
the operation, yet such an occurrence is not unknown. As
many as thirty or forty lambs have sometimes been found
dead next morning on a single hirsel; but, as a rule, the
death-rate is not more than one in every five hundred operated upon. Drinking whisky the night previous to operating upon the lambs has been credited as a source of evil at
cutting time, and so has indulgence in the pipe; but these
habits on the part of the operator have really no influence
in the matter. A man's breath must be in an abominable
condition before such a thing could possibly happen. The
more probable causes of ill-luck, as it is termed, are overheating the lambs previous to " cutting," and perhaps carelessness in performing the operation. When these precautions are taken, it is seldom that any casualty occurs.

Before proceeding with the operation of cutting, the best
of the tup lambs intended for sale or breeding purposes are
selected and put into a pen by themselves, where the merits
of each are subjected to the closest scrutiny. Their wool
is examined, and so is the colour of the face; and if their
symmetry and general appearance is satisfactory, after hear-
ing what the shepherd has to say regarding this and that
one's breeding, they are retained. Lambs, like all other
young stock, are more difficult to judge than old sheep.
Those in the best condition have usually the most attractive
appearance, but fat is apt to mislead the inexperienced as to
what may ultimately prove the best sheep. Although a lamb
may be small and thin in condition, it may possibly possess
the qualities essential for future development; and on that
account the whole flock should be very carefully scrutinised
in making the selections.

Cutting is then commenced with. The lambs are run
into a small pen suitable for catching them. The persons
required for the operation consist of one man to handle
the knife, another to hold the lamb, and either one or two

to catch. The operator and his assistant take up their position outside the catching-pen. The lambs are then caught singly and handed to the man whose duty it is to hold them up to the operator. The holder catches the lamb by taking a fore and hind-leg in each hand, and placing the back of the lamb against his left shoulder, he exposes the scrotum to the operator. It is necessary to hold the lamb very securely, and only strong men should be deputed for this work. A tall man is also better than a short one, as the facilities for cutting are much improved by holding the lamb at a considerable height from the ground. The operator is provided with a very sharp knife, which he holds in his right hand, and with the left he takes the scrotum between the fingers and thumb. Pressing the testicles gently upwards, he cuts about an inch of the purse clean off by one quick stroke of the knife. Taking both hands, he then springs the testicles from their socket, and, while holding them firmly at the root, to prevent tearing the flesh or fat, he draws them out singly with his teeth. It is unusual to put anything into the wound in the shape of salves. The purse is merely gently drawn into position, and a slight pull given to the tail, which assists in bringing the muscles into their proper position. With other sheep the tail is cut off at this time, but blackfaces do not require to have their tails shortened. A count is made of all lambs at cutting-time, and the numbers of both ewe and wedder lambs entered in the stock-book. The flock is then turned out to their pasture. The shepherd disturbs the lambs as little as possible for the next few days, merely keeping a close watch for any that appear ill, which he endeavours as best he can to revive. A solution of carbolic acid and oil is a good thing to allay inflammation, and a few drops poured into the purse will have a most beneficial effect. The exact number of both ewes and wedder lambs is always taken by the flockmaster at the time of castrating and docking. As a

rule, the two sexes are about equally proportioned.   Lord
Somerville found 101 ram lambs to every 100 ewe lambs.
Darwin has recorded that of 50,685 lambs which came
under his observation at one time, there were 98 rams to
every 100 ewes.

CHAPTER XVI.

SHEEP WASHING.

THE great object to be obtained in washing wool is not only to make it white but to make it bright. If too little yolk is left in the wool it will be wanting in softness; if too much it will become sticky, and after a time turn yellow. According to the late Dr. Voelcker, raw wool contains 42 to 50 per cent. of pure wool fibre, 20 to 22 per cent. of yolk (soluble in water), 7 to 10 per cent. of fatty matter, and 10 to 18 per cent. of moisture. There exists in some parts a long established, but more or less unequal and unjust, "rule of thumb," enforced by wool buyers, by virtue of which unwashed fleeces are subjected to a deduction of one-third of their weight when sold at the same price as washed wool. The assumption that one-third of the weight of a fleece is lost by washing it is, however, not warranted by the facts. If there is no clay or dung adhering to the fleece, all that it can lose by washing is the soluble yolk, varying from 20 to 22 per cent.; and this only if the sheep is shorn immediately, or as soon as the wool dries after clipping. It has been proved over and over again that when the yolk was allowed to rise again after washing, and the clipping done at the right stage, the loss by shrinkage did not exceed on the average 5 per cent. on the weight of wool.

The use of either hot water or soap is injurious to the wool. To cold-water washing there can be no such objection. For hill flocks, and, indeed, for all sheep that are confined on grass land all the year round, there is no better way than pool-washing—giving them two swims. For arable land sheep, tub-washing in cold water is better. Tub-washing has another recommendation where there is not a washing-pool on the pasture. Often in May and June, when you take sheep to wash, the roads are very dusty; and driving them along such roads after washing, they get dirtier than before, whereas in tub-washing the sheep do not leave the farm, and mostly miss any driving on the highway roads.

It is true that the percentage of washed wool from the colonies is decreasing, but there may be reasons for that which will not bear examination from the standpoint of sheep farmers in this country. The colonial sheep are chiefly merinoes, and the wool of these sheep is far too greasy to wash well in cold water; besides which the grease in a merino fleece is really worth something for soap and candle-making, and it would be wasteful to throw it away, although the cost of cleansing it at a subsequent stage is greater than the actual value of the residue. Many of the American wool-growers are also giving up the practice of washing the sheep; but as many of the American sheep are of the fine-woolled kind, it is needful to keep this fact in view when discussing the question on its general merits. With long-woolled sheep washing must be as necessary in America as elsewhere; and sooner or later the wool-buyers will find out the painstaking ones amongst their customers, and appreciate them accordingly.

In the Highlands of Scotland many of the farmers smear their sheep with greasy substances, to act as a winter protection, and on that account the wool is rendered difficult to wash. But an equal obstacle to the washing of Highland flocks is the difficulty of gathering the sheep on wild and elevated ranges for the purpose, and then bringing them together again ten days later for shearing, which last

I

operation must necessarily be done, even though the other has to be left undone.

The principal reasons alleged against sheep - washing are—

1. That it is injurious to the sheep.
2. That the washing can be done cheaper after the wool is off the sheep than before ; and,
3. That buyers will not pay enough more for washed fleeces to pay the cost of washing, together with the shrinkage in the weight of the wool, and that, therefore, the slovenly man in this is the gainer.

1. To talk about the sheep being injured by a dip or swim in the water during the month of June is simple nonsense. If the water is too cold for them, then what should be said about the barbarity of dipping sheep in cold water in October or November, and again in January or February, as is the rule with thousands of flocks? Numerous instances could be given where this winter dipping has laid the foundation of subsequent disaster to almost the whole flock ; but it would be difficult to prove that sheep-washing in June, with ordinary precautions, has ever been attended with ill effects to the sheep. In the case of the Highland flocks above referred to, injury might indirectly attend their being washed ; not from the mere washing or contact with the water, however, but from the injury they would be likely to sustain in being gathered in from the hills twice over within ten days—first for washing, and then for shearing purposes. On the other hand, it is scarcely necessary to tell practical men that sheep enjoy better health after washing. It will invariably be noticed that the lambs thrive and grow better after the ewes have been washed than they do during any other month in the year. A washed fleece is also easier to shear and easier to do up than an unwashed one.

2. In this country, on store farms at least, it is seldom

that any extra expense is incurred by sheep-washing. The shepherds do the work themselves, and their wages are the same whether the washing is done or not. Over and above this, wool is never so well washed, and never with so little injury to the fibre, as when it is washed on the sheep's back. The increased severity required to get dirty wool clean in the process of manufacture is injurious, and consequently reduces its value.

3. The American experience of wool buyers does not fortunately hold good in this country, for washed wool is here worth quite 1d. per pound more than the unwashed wool of the same quality; and so long as wool buyers act up to this distinction, as they now do, it is the interest of the flock-master to wash his sheep. Indeed he has every inducement to do so, for if any one is mulcted over this business it is the man who does not wash his sheep. It is a custom in many markets to pay the same price for unwashed as for washed wool; but where this is done it is also customary to deduct one-third of the weight when unwashed wool is bought, on the assumption that this proportion of the weight is lost in washing. This assumption is, however, altogether unjustified, because, as already shown, if the yolk is allowed to rise in the wool after washing before shearing the actual loss of weight need not exceed five per cent. With this fact before us, is it not clear that the grower who sells wool in the grease is unthrifty? Let us figure it out. Even tub-washing of park sheep should not cost more than $\frac{1}{2}$d. per head, and pool-washing does not cost the half of this, while the washing adds fully 1d. per pound to the selling price of the wool, and the difference between the weight of washed and unwashed fleeces is not more than five per cent. Now suppose an unwashed fleece weighs 6 pound and is worth 7d. per pound, its total value will be 3s. 6d.; but the same fleece if washed would be worth 8d. per pound, and show a gross value of 4s. Therefore, after

deducting five per cent. for loss of weight and paying $\frac{1}{2}$d. for cost of washing, the net value of the fleece would be 3s. 9$\frac{1}{2}$d., or a balance of 3$\frac{1}{4}$d. per sheep in favour of washing. This is tantamount to adding 7$\frac{3}{4}$ per cent. to the value of the wool clip.

In this estimate the cost of washing might very properly have been excluded, for all the labour bestowed in washing is saved in shearing, a washed sheep being far easier clipped than an unwashed one, and if the unwashed wool is exceptionally dirty, the difference in price will be more than 1d. per lb. The necessity for washing is not of course equally great in all cases. Long-woolled sheep and sheep that have been folded need it more than short-woolled sheep and sheep that are all their time on clean pasture. We must keep these facts in view when looking at the question as a whole, and allow that "circumstances may alter cases." But it is necessary to be clear on this point. It is wool and not dirt that is the marketable commodity in question. A good many clips of wool in this country are still sold by character, going year after year to the same buyers, who, having had them once or twice, will afterwards bid for those clips without seeing them—a very unlikely thing if selling wool with dirt in it ever becomes the fashion where the flocks have been folded.

Let us also make sure that those who complain of not getting as much more price for washed wool as will pay for the work and make up for loss of weight, really washed their sheep in the way they ought to have done. Washing is an important work, and as it is well or poorly done, so will be the reward. However carefully you may breed your sheep, and however superior the wool may be which they grow, your returns will be disappointing if the washing is badly or improperly done.

In a circular recently sent out by the Hexham Wool Merchants' Association they say: "The fleece should be properly washed or left unwashed, badly washed wool being of no more value than unwashed." Was the wool in the cases above mentioned clean washed? or was it smudged by driving the sheep along some dirty road as they came from the washing pool. And was the necessary time allowed to elapse between washing and shearing for the yolk to rise again in the wool? The yolk comes from the skin, and therefore if it is up again before shearing the alleged losses of weight by washing are to a great extent pure fiction. It is then only the dirt that is gone, and the wool staplers do not pay for that.

The number of days that should intervene between washing and shearing must depend partly upon the state of the weather, as well as upon the condition of the fleece. Yolk will rise quicker in fat sheep than in poor ones, but from seven to ten days is generally sufficient. The sheep should be properly docked previous to washing; thus preventing the dung and lumps of soil which adhere to the ends of the staples from discolouring the wool. Much of the advantage of good washing is frequently thrown away by carelessness in allowing the sheep to run after washing upon seeds or ploughed land, and also by allowing too long a time to elapse between washing and clipping.

CHAPTER XVII.

CLIPPING.

THE "clipping" is always a great event on a sheep-farm. According to the character of the season, the clipping takes place early or late, but as a rule blackfaced sheep are shorn some time between the 25th of June and the 10th of July. The neighbouring herds on perhaps eight or a dozen farms assist each other with this work. A day is fixed upon, and a start made at one of the farms within the circle of those included. This flock is then shorn, after which a move is made to another, and so on until the whole have been gone over. The advantage of plenty assistance on these occasions is because it entails fewer gatherings of the flock. The number of hands spoken for or advised to attend are generally sufficient to shear a whole hirsel in one day. Thus only one gathering of the sheep is required, and during unsettled weather it also saves a good deal of expense in the shape of providing for idle shearers, who would otherwise necessarily have been engaged to perform the work.

The success of the operation depends solely upon the weather. The day must necessarily be dry, and the wool on the sheep's back in the same condition. It is frequently very hot at the time of shearing, but sometimes it is showery, and the work is greatly interrupted. A decided wet day nobody minds; the clipping is then postponed for certain. It is when the weather is just sufficiently dry to induce the shearers to assemble and the gathering of the flock to be proceeded with that displeases most. The shearing may not have been commenced for more than an hour, when lo! a shower puts a stop to everything. The sky again clears, and hopes are expressed that it may be a good day after all. The sheep are taken out into the breeze to dry as quickly as possible. Another start is made, and no sooner than down falls another shower. This being repeated two or three times, it eventually becomes provoking, and it is truly ludicrous to hear the remarks made concerning the weather. But it is no joke to the farmer or the shepherd whose sheep are being clipped.

As the shepherds assemble from all parts, they assist with the gathering of the sheep nearest to where each resides. After the sheep have been folded on the clipping morning, the lambs are drawn off and turned to the hill, to prevent them being injured. The next thing to be done is to separate the hoggs and wedders from the ewes, as the wool of the former brings a higher price than that of the latter. A count is then made of all the flock, which the farmer duly enters in his book. During the "sorting" the herds that have come to assist keep a sharp look-out for sheep strayed from the "walks" they look after, putting such aside to take home with them when they return. After these preliminaries, the herds, to the number of from six to fifteen, seated on divot and wooden stools, or on the grass, now set themselves to the work of the day in earnest; and on an auspicious day, as the sheep-shearing progresses, the scene presented is lively and pleasant. Young lads are passing ewes as fast as required, and every shearer is busy. The professional skill of each operator is also something to look at. How deftly and cunningly the bright shears of the clippers do their work! Each shepherd bends to his task with skill and wisdom, and soon the cry of "tar" is

heard ringing on every side.   The following are recipes
for waterproof colouring or branding ink :—

(1.) Boiling tar and lamp-black.
(2.) Shellac, 2 oz. ; borax, 2 oz ; water, 25 oz. ; gum-arabic, 2 oz. ;
lamp-black, sufficient.

Boil the borax and shellac in water till they are dissolved,
and withdraw from the fire.   When the solution has become
cold, complete with 25 ounces water, and add lamp-black
enough to bring the preparation to a suitable consistency.
When it is to be used with a stencil, it must be made
thicker than when it is applied with a brush.

The above gives a black ink.   For red ink substitute
Venetian red for lamp-black ; for blue, ultramarine ; and for
green, a mixture of ultramarine and chrome yellow.

To the practised observer it is plain that the stock thus
passing through hands of the shearers varies much in value
and quality.   "Beasts are like bodies," we are told, "an' a
weirdless beast ne'er does weel."   A skilful shepherd likes
"a gude ewe" to handle, and his remarks meanwhile are
interesting to study.   "That ane's been o'er lang in the
moss, Jamie," we hear remarked ; and a "short-rise clip"
is not so prized as one that "fills" the shears.   Occasionally a "weardie" or " tinkler " is brought forward, and is,
consequently, shunned by one and all as being hardly worth
the trouble of operating on.   An interesting part of the
work is in watching how skilfully the shepherd handles his
ewe under the shears.   The very beast seems to know the
charm, and is quiet and obedient under it.   So soon, however, as it is shorn and branded with its owner's mark, then
the sheep feels its liberty, and bounds off in its nakedness.
It is amusing then to watch the lambs, and their difficulty
in again discovering their mother.   Between each other the
first sniffing and smelling that is seen appear odd and
strange.   In little more than an hour, after a day's clipping

has been finished, nearly all the lambs are again mated and off to the hills.

The buchts where the operation is performed are generally placed above the dyke that divides the highland from the lowland portion of the farm ; the hills around are clad in purple heather, and are studded here and there with the green bracken and the grey rock ; King Sol is looking down on all with his most genial smiles, and in his rays the shears so deftly handled by the herds gleam and glitter. All goes on merrily, the bleating of the lambs turned to the hills, and the response from their mothers in the buchts, alone casting a shadow of melancholy over the bustle. Being somewhat of a social and gala day, it is customary for the farmer to invite his neighbours to this handling ; and the approach of the jovial farmer or herd laddie, as he makes his appearance over the adjacent knowe, is heralded by a volley from the collies that would almost drown a volley from as many Enfields, and, as if for variety, disturbs the monotony of the bleating of the sheep and the snacking of the shears ; every newcomer has to be introduced in this uproarious and hostile fashion by the dogs. As the work proceeds standard anecdotes have again to be rehearsed, and the enjoyment and mirth they provoke is as fresh as it was when they were told for the first time by the same party at the same clipping some twenty years ago. The shepherd of the flock "buists," and is running about everywhere superintending, and occasionally going round with the bottle, for no sheep-shearing is complete without this time-honoured institution. The fleeces, as they are shorn from the backs of the sheep, are being rolled up and packed into sheets, ready for export. Perhaps the farmer himself may be clipping, or holding the gate of the "gruppin' bucht ; " but more likely basking in the sun, and entertaining his neighbours with a friendly chat on the prospects of the price of wool and lambs.

To the casual observer of a skilful workman the process
of sheep-shearing seems simple and easy of acquirement.
Let the aspiring novice essay the work himself, and his con-
clusions will suddenly reach the opposite extreme. The
ingenuity of man has wrought great improvement in the
appliances for shearing, and notwithstanding the changes in
the surface and covering, there is little room for doubt that
better work is done to-day than the old-time shepherds
deemed necessary or possible. It must be admitted, however, that much of the shearing is still performed in a
slovenly manner. Men who have worked at it for years,
and acquired skill in handling the shears, have failed to give
attention to certain minor details influencing the value of the
fleeces, and the comfort of animals as well. The standard
of good shearing is commonly mistaken. The average
beginner seeks for speed in turning off his work, when
really speed is secondary in importance to a number of
considerations—such an evenness of work, cutting the skin,
tearing the fleece, and injuring the sheep.

In becoming a successful sheep-shearer, the first and
most important essential to success is the position in which
the animal is held. When this difficulty is overcome, practice, with a determination to do the work properly, will
bring success. A sheep not properly held will be uncom-
fortable and restless, will kick its fleece into confusion, and
get many cuts and knocks into the bargain, entailing upon
the shearer greatly increased labour, and rendering it impossible for him to do satisfactory work. In the south of
Scotland the shepherds invariably tie together with a band
the four legs of the sheep, but the Highlandmen seldom
adopt this practice. They shear with the feet loose, and
it must be said this method is both quicker and easier upon
the sheep. The details of shearing can best be acquired
by practice under the teachings of an experienced workman—one who will show the learner just how and where

to open the fleece, and the proper position of the shears while so doing; how to avoid cutting the skin on the one extreme, or leaving too much wool on the other; how to keep the fleece during the progress of the work so as to preserve its form for convenience in rolling and tying and assorting into its several grades when passing into the hands of the manufacturer; how to keep the shears in order, and how to become proficient in the other details which go to make up the accomplishments of the expert shearer. Among sheep-shearers, as among men in every other calling, there are many good workmen, and many who are the reverse.

Sheep-farmers may be interested to learn that an Australian has invented a machine for shearing sheep, an exhibition of which was given at Euroko Station, near Walgett, on 26th November last. The implement is said to be the result of years of study and experiment. A large number of squatters were present at the trial, which, an Adelaide correspondent says, was highly successful. A number of sheep were shorn to the satisfaction of the judges, the time occupied being from three to four and a half minutes per sheep. Each animal was turned out as bare as a clipped horse, and without the slightest injury. Several sheep were shorn with the ordinary hand-shears and afterwards clipped by the machine, when an additional $\frac{1}{2}$ lb. to $\frac{3}{4}$ lb. per sheep was taken off. It is also claimed for the new machine that the wool removed by it is more uniformly cut, there being no second lengths. If this be so, then the wool would not only be more speedily removed, but the clip would be both heavier and of better quality when cut by the machine. Many inventors have spent much time and great pains in trying to perfect a sheep-shearing machine. Should this latest attempt prove to be a success, its importance to sheep farmers at home as well as in Australia will be very considerable.

CHAPTER XVIII.

DIPPING.

DIPPING is considered a necessary dispensation in all well regulated flocks. In former times flockmasters had recourse to salving or smearing as a preventive and remedy for parasitic attacks on their sheep. That system, however, has now been laid aside in favour of the more effective and less expensive process of dipping. The old method of smearing—an operation which consisted in soiling the back and sides of the sheep with a mixture of tar and grease—had the reputation of greatly damaging the selling value of the wool as well as being somewhat costly. For these reasons, and also owing to the rise in the price of wool, farmers were not slow in discarding the practice of smearing, and no one has ever greatly mourned the change. It was laborious work, and besides so long as it was practised the country was never free of scabbed sheep. Neither was it so good a destroyer of keds and ticks as could be wished; so that on the whole we may be thankful for the discovery of improved methods in keeping the sheep free from the ravages of vermin and disease.

Sheep of every breed, and in all countries, are liable to become affected with parasites—described in another portion of this book—and it is therefore necessary for their welfare to keep them as clean as possible. It would be contrary to the laws of nature to expect them to thrive and prosper otherwise. The question then arises as to the proper season and method of accomplishing this very desirable purpose. Under ordinary circumstances, dipping once a year will have a beneficial effect, although to ensure complete success the operation should be performed twice a year if possible. The season of the year most favourable for the destruction of the ked is sometime in the autumn. At that period of the year the insects are, for the most part, fully developed, and therefore more easily killed. The hard skin in which the ova of the ked are enclosed, renders their destruction in that state both difficult and uncertain. But as the most of the eggs are laid and hatched during the summer season, the full benefit of dipping in the autumn is then secured.

It is not, however, to be supposed that the autumn is the only period of the year suitable for dipping. As a matter of fact, a great many flock-masters prefer to perform the operation at other seasons. It is not advisable, for the sake of the health of the sheep, to dip them during cold weather; and it is also a mistake to subject ewes heavy in lamb to the unavoidable rough usage occasioned by the process. For these reasons, dipping should either be performed shortly after the shearing time or about October, before the cold weather of winter sets in, and ewes in lamb not later than January or February. After the sheep have been shorn is a time preferred by many for dipping. The wool is then very short, and less material suffices for soaking the fleece, besides the eggs are more liable to meet with destruction at that time. It is also a fact that the keds, not preferring a bare skin, as it affords them insufficient shelter, manage to find their way from the shorn sheep on to the rougher coats of the lambs. So generally is this the case, that some farmers dip the lambs only at this time, and while killing most of the keds, do the lambs a great amount of good besides. Dipping the lambs has, in recent years, become a very common practice, at least on farms where the sheep can be readily collected, as it greatly improves their selling appearance. Dipping the lambs only, no doubt exterminates the greater part of the vermin for the time being, but the ova which are left in the wool on the old sheep soon become active, and as numerous as before; so that to thoroughly eradicate the pests both the old and young sheep should be dipped at the same time. The best results of all are unquestionably attained by having the operation per-formed twice during the year—first after the sheep have been shorn, and again either a week or so before the rams are put out in the autumn or about the beginning of February. The summer dipping is useful in preventing the annoyance of flies, and it also materially promotes the growth of wool, indirectly of course, owing to the sheep being better thriven. The winter dip, if consisting of the proper materials, may also be made to act as a protection against wet and cold; and these are important advantages, while at the same time keeping the sheep almost entirely free of vermin.

The best kind of dip or composition to use for destroying sheep parasites is a problem that does not admit of any definite solution. As is well known, there are no end of patented " dips " in the market, all of which have their several good qualities to recommend them. It is not our intention, however, to say one word in praise of or against any of these preparations, as far as their respective merits or efficiency are concerned. We believe them to be generally suitable for the purposes for which they are intended; only, in the interests of the farmer, we are bound to state that a home-made dip can be prepared at a much cheaper rate, and one which, if properly compounded, will prove equally effective. Those who prefer to purchase the ready-made article must choose their own merchant—it would be unfair to give any of the manufacturers a preference before others, who, for anything we know, may sell an equally good article; only, not having tested them all, we could not speak from experience. Personally we have used the preparations of perhaps half-a-dozen different makers, and can safely say that when the directions for use were strictly adhered to, the results were highly satisfactory. The cost of these varied from 4s. to 8s. per 100 sheep.

Any farmer may, if he chooses, compound his own dipping composition, the ingredients of which are easily obtainable at almost any chemist's shop. With this object in view, the following recipes for home-made dips are given; and it may be mentioned that the most of them are the actual preparations in use amongst farmers throughout various parts of the country, and are prepared simply by dissolving in boiling water, and then added to 80 gallons of cold water in the tub. The cost of either will run from 2s. to 4s. per 100 sheep :—

No. 1.
1 gall. carbolic acid,
2 lbs. arsenic,
4 lbs. soda.

No. 2.
2 lbs. arsenic,
1 gall. pitch oil.

No. 3.
1 gall. pitch oil,
4 lbs. soft soap,
3 lbs. soda.

No. 4.
1 gall. carbolic acid,
2½ lbs. arsenic,
1½ galls. spirits of tar.

No. 5.
1 gall. carbolic acid,
2 lbs. arsenic,
3 lbs. soda,
3 lbs. soft soap.

For a summer dip no arsenic should be used, as it is apt to injure the lambs. At any time arsenic is a dangerous article to use too freely, and on no account should more than the stated quantities be applied. Some flockmasters have condemned the use of arsenic in any form, as they have experienced serious loss therefrom, while, on the other hand, others declare that no dip that does not contain arsenic in some form was ever found to be satisfactory.

The so-called non-poisonous dips are not strictly what they profess to be, as no composition will kill keds promptly and surely unless it contains a poison of some sort; and arsenic, if properly and carefully handled, is quite as safe as any ingredient for that purpose.

In selecting a dip, farmers should also bear in mind that many compositions in use are sadly injurious to the selling value of the wool, on account of the materials used not being susceptible to washing out again. The Messrs. Greig, wool-brokers, Leith, in their latest circular to the trade, give the following advice :—" We beg to draw our customers' attention to the fact that some dips used for blackfaced sheep *will not wash out of the wool*, and spinners complain that no amount of scouring will make the wool white. We would recommend that for dipping only white grease or butter should be used, which will scour quite out; otherwise farmers will be disappointed with the prices obtained."

A much disputed point among hill sheep farmers, is whether grease should form one of the ingredients of a dip. It is usually supposed that grease applied to the wool in almost any form, acts as a sort of water-resisting agent, and upon this theory many farmers add a quantity of oil to their dipping compounds, with the idea of rendering the fleece waterproof, or impervious to the rain. Other equally experienced men, however, assert that to grease the wool, if not decidedly injurious, is productive of no good effect: that instead of the rain being shielded off, and the sheep kept dry and warm, the grease "clogs" the wool, forming it into masses and creating openings in the fleece, and allows the rain to enter direct to the skin. And, further, that when once the fleece is thoroughly sooked with wet, it is much longer in drying, which keeps the animal cold for a longer period, greatly to its hurt and discomfort. It is also a fact that greased wool, although heavier than the same ungreased, brings a proportionately smaller price in the market. This can be accounted for in several ways. In the first place, the oil in the fleece is of no material value to the manufacturer, and, consequently, the extra weight which it gives has merely to be deducted from the selling value. But the chief cause of the antipathy which brokers hold against the greased staple, is because British manufacturers have not the proper machinery in their mills for cleansing it. Very few farmers are perhaps aware that the greased wool sold in Scotland has to be sent to America to be cleaned before it can be used at home for weaving purposes. In America the bulk of the wool grown is from a naturally very oily-skinned breed. They have, therefore, suitable machinery in their mills for extracting the grease, a somewhat expensive process, which our home manufacturers have not yet adopted, and in consequence they either cannot or will not make use of that class of wools.

A study of the natural history and distribution of the sheep species clearly indicates that to artificially grease the wool is prejudicial to the health of the sheep. Every country, be it cold or hot, wet or dry, has in the lapse of time developed a kind of sheep peculiar to the geological and meteorological exigencies of the situation. And to distribute these breeds successfully throughout the world, there are certain natural laws that must be observed. In Spain, for instance, there has originated at some period or other a greasy-woolled breed of sheep called Merino. About the beginning of this century attempts were made to introduce the Merino into this country. The experiment failed, however, because it was soon discovered that a naturally greasy-skinned sheep could not thrive in our moist and colder climate. This is one of nature's lessons, which tells us that if we strive by artificial means to upset her teaching, we must either pay dearly for the experiment or fail in the attempt. It would be an equally easy matter for the Spaniards to adopt any of our Scotch or English breeds of

K

sheep in preference to their Merino, and it would have been done long ago had the change been at all desirable. As it has not been done, we may rest assured that each breed is best suited to its own country; and the inference is that, wherever the rainfall of a country is at all heavy, greasy-woolled sheep never prosper. They can stand cold, and also heat to almost any extent, as is evident by the success of Merinos in America, but they must be protected from wet. Now in Scotland, where the rainfall is comparatively heavy, is not artificial oiling of the fleece in direct opposition to the natural requirements of the animal? We unhesitatingly answer in the affirmative; and no amount of quibbling logic can disprove it. The damper the climate, the drier and warmer must be the animal's covering. Heat is the great element of life, and cold the chief destroyer; but the latter when accompanied with wet is doubly fatal. Grease or oil applied to the fleece of the sheep is a powerful conductor of cold, and it has the further defect of being a retainer of wet, than which there is no better promoter of cold and its attendant evils. The surprising part in connection with greasing sheep in Scotland as a winter protection is, that it should ever have been practised to the extent it has. Shrewd flockmasters, however, have now discovered that it is a false attempt to improve upon nature, as well as being an expensive usage without one single feature to recommend it.

Reverting to the practical work of dipping, the erection of a good, serviceable " dipping machine," is an important part of the operation. There are several kinds of these machines in use throughout the country, varying a good deal in construction as well as in cost of erection. Dipping is an operation which few care to spend too much time over, and it is therefore advisable to have the machine of a size suitable to the number of sheep on the farm. Tenant farmers have generally to provide their own dipping machines, and in that case the erection is usually made entirely of wood, as it is cheaper than the more durable and costly concrete. The concrete bath-tubs are, however, much more serviceable, and while the tenant may properly enough bear the cost of the necessary fittings, the landlord ought not to refuse to erect the bath proper or tub, which should be built of concrete, and remain a permanent fixture upon the farm. An ordinary swimming bath, suitable for a flock of say 1000 sheep, should be about 15 feet long, 24 inches wide, and 4 feet 6 inches deep. At each end of the bath a gangway is made to enable the sheep to walk both in and out of the tub. The entrance gangway is short and steep, and about 3 feet long; the exit gangway about 5 feet long, and not so steep. Thus the length of the swim is reduced 7 feet, and this part only requires to be the depth above stated. With a machine, as it is termed, this size, 1000 sheep can be readily dipped in a single day, with the assistance of four men and a boy. A good day with signs of dry weather to follow should always be chosen for dipping. The work should not be done carelessly or too hurriedly, and an intelligent person should take charge of the mixing of the dip.

CHAPTER XIX.

MOUNTAIN GRAZINGS.

THE mountain grazings of Scotland embrace almost every formation it is possible to mention. As a rule, however, the granite rocks claim the alpine pasture, the limestone and its associates the uplands, and the sandstone the low-lands. If we take these as representatives, and make allowances for geographical peculiarities of districts, we have set down pretty correctly a distinct base for the sheep-grounds of Britain. And, as has been well said, each of these is distinguished by striking characteristics besides that of the soils. Their very outlines, for example, at once indicate each, and, of course, their main features as to temperature, rainfall, and herbage, all being more or less regulated by climate.

Small as our island is, a day's journey on it is sufficient to traverse rocks of every geological age; and even the most careless observer cannot fail to be struck with the diversity of appearance presented by the surfaces of the different strata over which he travels. In a majority of cases the soil is derived from the subsoil; so that, for the most part, the soil indicates the rock. It is true that in regions where there is a thick cover of drift, the soil has little or no relation to the solid rocks below the drift. But this "drift" is really the surface-rock in the geological sense of the word; so that there is no exception there to the rule that soil is rotted subsoil, and subsoil is rotting rock.

Thus in any estimate of the fertility of land the nature of the underlying rock comes into consideration; for not only the depth, but the texture and composition of the soil depend to a considerable extent on the rock beneath, and its productiveness is dependent on these. Thus soils formed from rocks which abound in phosphates are often of extraordinary fertility. Even the fossils and shells that are found in, and are characteristic of, rocks increase the value of the soils where they occur. The natural vegetation of any locality is so entirely dependent upon the nature of the soil that the geologist often receives great assistance in mapping the boundary-lines between the different strata from careful observations of the plants which grow there. To such an extent, indeed, is vegetation influenced by soil and climate, that the experienced farmer can form an opinion as to the fertility of the land from the species of plants growing upon it. But even a naturally fertile soil is frequently rendered unproductive by deficient water-supply or defective drainage.

It is not to be supposed, however, that the differences in agricultural features are due to the soil alone, for climate has as much influence as soil in modifying agricultural operations; but since climate is regulated mainly by contour, and contour depends chiefly upon geological structure, there is the same primary cause for variations in both cases. Nor must it be forgotten that it has been the tendency of former systems of agriculture to overcome all obstacles to the growth of crops, even on soils which are naturally unsuited to them; and thus the natural diversity of character which formerly existed is, to a great extent, destroyed. But in these cases the improvement is artificial, and the natural conditions would reappear if constant attention were not paid to the maintenance of the soil in an improved state. To the ordinary observer the appearance of a naturally fertile soil, and of a barren soil rendered artificially productive, may be very much the same; but to the farmer the difference is extreme, for the latter can only be made to yield good results by endless trouble and expense. Hence, not only do the natural capabilities of the soil vary with the geological structure of the rocks from which it is derived, but facilities for improvement and the amount of attention necessary to prevent deterioration are similarly influenced.

Soils suitable for Sheep.—Sheep are found to thrive well on almost every variety of soil for a time, but on some they deteriorate sooner than on others. Not only has the quality of the soil an important influence on them; the physical structure of it has also to be considered. On certain stiff clay soils, popularly called " sour," they will soon lose weight, and probably also decline in health; while on a soil which is much inferior to the other in respect of the abundance of plant-food to be found in it, but which is dry and porous, they will thrive very well. On soils resting on strata of granite, clay-slate, or other allied formations, and which have been formed chiefly from the disintegration and decomposition of these rocks, they will seldom fail to thrive. On soils where the bracken and other ferns grow abundantly sheep find their favourite pasturage; also where the great variety of plants grow into the constitution of which potash largely enters. In connection with sheep thriving well where there is an abundance of potash, as indicated by the prevalence of certain plants, it has to be remembered that the fibre of wool is greatly dependent for its protection against rain and inclement weather upon the emollient substance called " yolk," which keeps it perfectly soft and flexible, and renders it waterproof, and that this substance contains potash in considerable quantity, while the wool itself contains sulphur, both of which must, of course, have been derived from the soil. When, however, we put upon the soil an animal which feeds upon the plants, and retains for the upbuilding of its tissues some of the constituents which the plants have derived from the soil, we see at a glance that, unless the amount of these constituents be maintained in a soluble or assimilable form, by the continued disintegration and decomposition of the rock at a rate equal to that at which the animal is removing it, the ultimate exhaustion of these constituents may be easily calculated on.

In old and deep fertile soils there may be abundance of all the needed food of plants, but the constant depletion of one of these out of proportion to the others will in time affect the plant-growth, and through it ultimately the animals feeding thereon. The character of the soil affects the food and the water, either or both, and brings the system into conditions that form a class of ailments that no care or hardiness can withstand. It is only from long acquaintance that we can tell whether the soil is healthy or otherwise. The most experienced farmer could not judge it merely from observation, and much less are we able to describe what is good or bad with the pen. In certain districts apparently excellent sheep-lands are sometimes very unhealthy; and, again, in others the poorest imaginable looking farms are sound and healthy, although in the latter the general quality of the sheep is much inferior to those of the former. Experiments and careful observations might overcome the local deficiencies or causes of trouble. It might be the presence of mineral in the water, which an antidote would remedy. It might be from some parasitic trouble, as the liver-fluke, or it might be an unhealthy condition of the soil, which is continually introduced into the system of the sheep by the food. These could all be counteracted if we knew what it was, and what would be the antidote.

Soils and Grasses.—There is no class of farms so variable in character and value as hill grazings. It is the land that makes the sheep; and on the nature of the soil, on the

kinds and varieties of the natural grasses, on the conditions
of climate, may be said to depend the weight and quality of
both sheep and wool, also the liability of sheep to disease,
and whether food will be abundant or scarce at the proper
seasons. The exposure or lay of the farm is not of less
consequence, for some grazings are far more easily stormed
than others, and it is a great advantage when the sheep are
well sheltered from cold winds.

Large areas of the hills of Scotland consist of clay, a
deposit of the glacial period. This description of soil is
often found on the lower slopes of hills, on slightly undu-
lating ledges, and on terraces or plains which frequently
relieve the rapid declivity of mountains. The herbage is
usually rank and nutritive, and is distinguished by a dark-
green colour. Throughout the year it affords the best of all
pasturage, and rears the largest sheep of any description of
soil. A large part of Border hills and the valleys of Sutherlandshire are of a clayey nature. The prevailing sheep of
those districts are Cheviots, and, strange to relate, the blackfaces are not found to thrive so well on that description of
land.

There is a great quantity of bog or marshy land on some
hills. It is not much fed upon in summer by sheep, but it
is of great use to them in winter. It abounds in spret and
finer grasses, which are fed by sheep in winter as well as
summer. The bogs in their abundance also yield hay for
winter foddering, a most essential provision on a hill farm,
especially if it lies in a high district. Then again, on ground
of this class, there are fewer deaths, and fewer barren ewes
in the flock, the produce is greater, and fetches a relatively
higher price, and fewer lambs are required to keep up the
stock than on any other kind of hill pasture. It is a great
improvement to drain these bogs; it makes the grass more
wholesome, increases its quantity, and makes it eatable in
frosts, whereas in their natural state they become one sheet

of ice. These bogs are also improved by mowing every third year. When they are not mown they grow up into little hillocks, which the sheep will scarce touch. The hay made from them is also of great value in winter, and is preferred by the sheep to any other of foreign growth.

Peat or moss is a very common variety of soil. It makes but a light sheep, and is best when mixed with grassy land ; but sheep are not easily wintered on hills where there is no moss ground. It is found in all positions, from the lowest valleys to the highest mountain-tops. It is of most value when low down on the hills, for then the sheep can reach it more easily in stormy weather. In one place it yields short nutritive grasses, and in another rank, worthless heaths. Peat varies in character according to its depth and wetness, and to its deposition on plains, gentle slopes, or rapid declivities. On the level, and surcharged with water, rank heath and tufts of yellow fog (*hypua*) are its chief produce. When in a drier state, although in a similar position, it produces some of the earliest and most nutritive of spring plants, such as cotton grass in spring, deer hair in summer, and ling which grows all the year round. When deposited upon a gentle slope or upon a porous subsoil, and with a consistency and compactness of character, it generally produces short, rich green herbage, or a fine quality of heath, both of which make valuable pasture. Hills on which peat and clay alternate, yielding the cotton grass for spring food, rich green herbage for summer, and rough bog for winter, possess the most desirable qualities for a sheep-run.

" Bent " land also yields some excellent grasses of a coarse kind. The two chief varieties are stool-bent and white-bent. The former is an evergreen, and more valuable than the latter, which dies off in autumn. Fine dry pasture or "lea" land only affords summer grazing. It is entirely free of heaths, and yields no vernal grasses; consequently it is very barren in spring, is easily drouthed in summer, and liable to injury by naked frosts. In a moderately moist summer and mild winter the stock on such lands fare admirably; but under opposite conditions it goes hard with them, and they invariably suffer severely by the fluctuations of the weather. As a rule, stock on this variety of land require to be hand-fed in spring.

Heath grows on almost every kind of soil; but moss heather is best for sheep. Not only the portion of heather on the grazing should be seen to, but the proportion of old and young heather. Sheep, like birds, only eat heather in its young state, and the value of many moors is greatly lessened for sheep-farming by an undue proportion of old heather.

Loam or Grit may be regarded as a mixture of those soils which have been noticed. It varies in its character, which may be ascribed to the formation on which it rests and to its produce. Where sand forms its principal ingredient the herbage is coarse and tufty, and comparatively worthless for sheep-feeding; while, when it is of a light hazel colour, it yields a valuable sward of fine short pasturage, much relished by sheep in summer.

The remaining soils of Scotland partake more or less of the same character, varying in quality and texture according to formation, deposition, and altitude. The grasses are equally diverse, their whole value depending upon the conditions in which they are found to exist. Heath, for example, when growing upon a moderately dry mossy soil, affords one of the most valuable of all hill plants; whereas on a cold clay it grows so coarse as frequently to be of no value whatever. It is the same with all the other grasses, their nutritive value being subject to the soils on which they exist. The best hill farms are those having a proportionate extent of all the soils mentioned, producing a variety of pasture suitable for grazing at all seasons of the year.

A hill grazing, then, is valuable, more or less, according

to its proportion of these soils and pasture plants. A good summer pasture is, of course, an essential thing in any case; but where the pastures produce only summer grasses the flock is not so easily wintered, and the sheep, for the most part, come out poor in spring. A grazing with the soil and pasture all of one kind is seldom to be met with, and as seldom to be desired. Grassy or bog land alone, however, of itself, will grow good sheep, and keep them in condition throughout the year. The value of every variety of hill pasture depends a great deal on its being of a mixed character. The altitude of a farm has an important bearing upon its value, independently of either the soil or grass. Many of the highest mountains, although capable of producing an excellent bait for sheep in summer, are worthless as such during the winter. They are enveloped in snow for five or six months of the year; consequently the sheep have to be removed during that time, and their value must be gauged accordingly.

Healthy and Unhealthy Sheep Land.—In determining the kind of land upon which sheep are most free from disease, it is first necessary to distinguish the elements needful to the development of the animal system. It is known by chemical analysis that wool, skin, hair, and horn are closely allied in their composition. They are nitrogenous compounds, composed of the same elements, and in nearly the same proportions, being compounds of carbon, hydrogen, nitrogen, oxygen, and sulphur, with minute portions of earthy matter, which give firmness to the texture. In addition to these there are other ingredients of which the whole carcass is composed. These are potash, soda, lime, magnesia, oxide of iron, chlorine, silica, and phosphoric acid, all of which are required to maintain sheep in a healthy state of existence. Land or soil that contains these elements in sufficient quantity, neither too much of one nor too little of another, will doubtless afford the most healthy and invigorating sheep pasturage. The plant is the medium through which all the elements of soil are carried to the sheep, or, as it has been tersely put, " the dead earth and the living animal are but links of the same chain of existence, the plant being the connecting-bond by which they are tied together." Without entering into a detailed account of the composition of the various rock formations, it will be sufficient merely to notice their effects upon sheep generally, as confirmed by actual experience.

On the granite ranges the disease called " vanquish " or " vinkish " is invariably met with. This disorder may be interpreted as a wasting away, and when sheep are first affected with it they become watery about the eyes, lose their middle, assume a sickly appearance, and if not removed to some other formation will eventually succumb. If shepherds, when going their rounds, instead of quietly passing by ailing sheep, would drive them off to some other part of the farm, which they knew to be different in character, or failing that, to some other district, they would have a great deal fewer skins to count during the course of the year. When shifting sheep, however, for this cause, it is necessary to take them to an entirely different formation. The change on to "Old Red," "Carboniferous," or more especially the trap, better known as " whinstone," acts like magic, and if not too far gone they soon recover.

A combination of the granite and trap formations is considered to be even more subject to " vanquish " than the pure granite. This is supposed to arise from the alkalies contained in the granite which frequently overlies the other being washed from the higher ground on to the lower ; and as both of these formations contain a large quantity of potash and soda, when the two soils are combined the alkalies are then present in abundance, and consequently act injuriously upon the health of the sheep. In certain localities, however, it has been asserted that where the detritus of the granite is washed on to the trap the death-rate is not unusually heavy. Perhaps sufficient care has not been taken to discriminate between the two in these cases. As has been conclusively ascertained, the granite-trap formation is decidedly subject to this complaint—so much so, in fact, that sheep cannot be maintained upon land of this kind for any considerable period at a time.

The trap formations are, perhaps, the most healthy for sheep of any found in Great Britain. And well it is so, as they embrace the largest area of all our hill grazings. The greater part of Galloway, Dumfries, Peebles, and Selkirk, and a considerable area of Ayr, Lanark, Edinburgh, Roxburgh, and Haddington are on this formation. In the north, although considerably metamorphosed, it is found over the greater part of Perth, Kincardine, Moray, Nairn, Banff, Inverness, Argyle, and Sutherland. Disease is not more common on the northern belt of this formation than it is on the southern; but the death-rate is nevertheless higher in the north, owing to exposure and an altogether higher elevation.

There are many other causes which foster disease and increase the death-rate apart from geological formation. Overstocking, bad management, and exposure to certain winds are sometimes the most prevalent causes of fatal disorders. Diseases such as louping ill and braxy are common to trap lands, and to some others also; but whether they are due to soil influences has not yet been ascertained, so that, in the meantime, these sheep complaints cannot be associated with any particular formation.

The Carboniferous formation prevails to a considerable extent throughout the counties of Fife, Edinburgh, Lanark, Ayr, Renfrew, Dumbarton, and the north of England. On this formation the best class of blackfaced sheep are found to exist. The mountain pastures of these districts, as a

rule, consist of grassy slopes, with poor bog land on the
bottoms, and a mixture of heath and moss on the hill-tops.
Compared with the trap and granite, the Carboniferous soils
are considered a trifle less healthy; but the formations are
frequently so interwoven together that it is difficult to determine which of them is really the soundest grazing. On the
limestone sheep usually thrive well, and they prefer the
natural herbage they find there to that of any surrounding
formation.

The Laurentian, Cambrian, Old Red Sandstone, and
Boulder Clay groups form so small and indistinct a share
of the mountain grazings throughout Scotland that very
little importance is attached to them. The two former
produce a small-sized sheep, while the latter carries a sheep
above the average in size. The sandstone is credited as being
a healthy kind of land for sheep, and the wool is said to
be about the best in Scotland. On neither of these groups,
however, are blackfaced sheep particularly noticeable. They
are more frequently possessed by the Cheviot breed.

Peat soils compose a large part of the sheep-walks of
Scotland. Peat arises from the accumulation of neglected
vegetable matter in moist situations. Where successive
generations of plants have grown upon a soil, unless part of
their produce has been carried off by man or consumed by
animals, the vegetable matter increases in such a proportion
that the soil approaches to a peat in its nature; and if in a
situation where it can receive water from a higher district,
it becomes spongy and unfit for the growth of the finer
varieties of plants.

Another mode in which peat is formed is by the gradual
accumulation and decomposition of aquatic plants in shallow
lakes and stagnant pools. This kind of peat is of a more
loose and spongy quality, and the fermentation which takes
place seems to be of a different kind, more gaseous matter
being evolved.

What has greatly contributed to the growth of peat is the destruction of ancient forests, either by the operation of some natural cause or by the hand of man. When water gets collected or choked up—as one finds it in a morass, for instance—an active vegetation soon begins to form a peat or vegetable soil, and the depth of the bog is only determined by the height of overflow of water from it. Many plants continue to grow in this, and by gradual decay form peat; especially the "bog moss," *Sphagnum*, which, while it grows above, decays beneath. It is singularly variable in character and value as grazing land.

A peaty soil grows a lightish-boned sheep and an inferior class of wool, the mineral elements requisite for these purposes being awanting. Foot-rot is common to this description of land, being superinduced by the excessive moisture, and also by the hoof not being worn down as fast as it grows.

The herbage upon all these different formations must of necessity vary in quality. Where all the elements best adapted for the growth of the plants are present in the soil, the sheep fed upon it are likewise in the same healthy condition; whereas the sheep fed upon grass which does not contain all the matters required are liable to be poor in quality and predisposed to disease. Speaking generally, the death-rate among hill-sheep from geological influences may be reckoned at from 3 to 5 per cent. per annum. Occasionally the loss will be much heavier, and sometimes it will be less, varying according to the seasons, which may be wet or dry, warm or cold; and the seasons have an all-powerful effect in themselves, without taking the soil into account at all.

Influence of Soil on Sheep.—The influence of soil plays an important part in the breeding of sheep. Unless the ground be adapted to the breed maintained, success will be difficult to achieve. The majority of the hills throughout Scotland are well suited for rearing blackfaced sheep, yet on the grassy hills, in the south and west, for instance, where that class of soil known as peat moss is comparatively scarce, they have not been found to thrive so well as in other parts where it is abundant.

If it were asked how all our breeds originated, we would answer, partly through selection and cultivation, but chiefly by the influence of soil and climate. It cannot be proved that breeds owe their origin to the influence of soils, but the fact may be reasonably assumed. Were a few sheep of the same breed to be turned on every kind of soil found in Great Britain, we know from experience that in a few years they would become so much altered in appearance and in outward form as to be unrecognisable as belonging to the same family. And whatever form or size they, after a time, naturally assumed could, by selection, be altered sufficiently to create a variety of breeds. The same effects are apparent in cattle and horses. Wherever any particular breed is found, they invariably correspond in size and weight with the nature of the soil in the same district. They may have been altered somewhat in colours, according to the tastes of the breeders, but the animals are in reality what the soil has made them.

Nature has certain laws which she compels us to obey, and none more strictly than in the matter of breeding live stock. Were it not for that limit which soil and climate put upon races of animals, new breeds might be evolved indefinitely. But every breed is in a manner indigenous to the soil, and all owe their originality to a distinct description of geological formation. That is a point which no breeder of live stock can afford to overlook when settling in a new locality, or in attempting the breeding of imported stock. In exceptional cases, where the soil and climate are nearly similar, or are improved upon, stock can be removed with safety and success from one district to another; but otherwise such experiments are bound to fail to the extent of the difference in natural conditions. And while geological influence is all-powerful in regulating the distribution of breeds, it plays an equally important part as regards the health of farm live stock. It has been ascertained, for example, that in a Carboniferous district certain diseases are prevalent which are markedly absent in the New Red Sand stone. This is attributed to the presence of iron in the one set of rocks, and its complete absence in the other. Again, the Silurian and other formations, which are notably deficient in lime, are subject to many diseases also peculiar to themselves. Nearly every district has its own individual experience in diseases, which are undoubtedly related to the prevailing geological systems, but, from want of research, remain a mystery to even our ablest scientists. Chemical analysis has revealed that every part of an animal is composed of certain elements found in the soil. Those who know exactly in what proportion these elements exist upon their farms we believe to be few indeed. But until such facts are explained by the science of geology and its relation to stock and crops, farmers must continue to grope in the dark. Enough has at least been ascertained on these points to encourage further investigation; and it is hoped that we may yet be able to map the area, and region endemic diseases as accurately as the different characters of the surface soils.

Deterioration of Hill Pastures.—It may be granted that the main elements of fertility are all contained in the soil itself, reproducible by the effects of the atmosphere and other agencies; and that if only the land be given time, and is not made to yield up its supplies of plant food faster than they are formed by natural means, the soil will retain its normal condition of fertility for an indefinite period. From the nature of hill pastures, they do not admit of much more

L

being carried off them annually than nature restores to them—as a whole, *i.e.;* but the evil is that the drain is often heaviest on those elements which are in the least supply, while it leaves in a measure untouched other elements that are at present in abundance. Thus lime is the most essential element in a soil devoted to the rearing of stock, yet it is singularly wanting in, and often altogether absent from, the rocks or soils on which our hill pastures chiefly rest. Of these rocks the Silurian give, perhaps, the widest range of hill pasture, occupying as they do most of the Lowland hills of Scotland, the greater part of Cumberland and Wales, and a large tract of the Irish mountains. The altered Silurian rocks (gneiss, schist, &c.) occupy almost the entire Highlands in the north of Scotland. Granite crops up among the metamorphic and Silurian rocks in Caithness, more largely in the Grampian range, a little in the Kirkcudbright hills, and to a considerable extent in several parts of Ireland. Trap occurs more sparingly than granite even, but it appears in the north-east of the Cheviots and in Peebles and Lanark hills, and to a greater extent in Antrim. Millstone grit and Carboniferous limestone occupy the hills of Northumberland, Lancashire, and Yorkshire. The trap, granite, metamorphic, and Silurian rocks are all hard and intractable, and form little soil, as they are difficult to decompose. Trap and Carboniferous limestone generally contain much lime, and the addition of it to them is indispensable to their high fertility.

The most fertile of soils have only a limited available stock of materials of supply, and where more than this is continuously taken away, without making compensation, the exhaustion of the soil is only a matter of time. When the process is slow it is no less sure; and although this wear-out of hill pastures is in most cases so inappreciable annually, from the extreme meagreness of their yield, yet the effects of it are not so indistinct that they cannot be traced. There is evidence of it in the alleged deterioration of the sheep on some of the hills, and in the fact that hill lambs at the beginning of the century were weaned about a month earlier than they are now; as also in stories of more sheep having been kept on many of the hills in former times—all of which assertions have at least a modicum of truth in them, after making every deduction for the improved sheep of modern times, the disuse of ewe-milkings, and other things.

The reproductive effect of the atmosphere, the annual vegetable decay, the action of worms, and the droppings of animals—which, in favourable circumstances, may together be far more than a set off for all that is removed from the pastures by any system of sheep-farming—is considerably lessened, if not rendered *nil*, on the hillsides by the rapid denudation and washing of the surface-soil which occurs in such situations. The atmosphere is indeed more an agent of destruction than of reproduction on soils of steep declivity. The heavy rains which fall on the hills carry many soluble substances away with them, so that the upper soil is kept poor from this cause alone; and good drainage will not prevent it on the slopes of the hills; while vegetable matter which serves to retain the soluble salts and keep them from being washed away is, as already remarked, most wanting in these very soils. The amount of mineral matter thus removed is very great. Professor Ramsay, taking the thickness and curvature of the Silurian rocks in the Woolhope district, says that to restore the section the same as the rocks were probably bent would give them the great additional height of 3500 feet. The annual rate of denudation is estimated by Professor Geekie at $\frac{1}{6000}$th of a foot from the general land surface of the country. This, of course, does not come equally from the whole area of drainage. Very little may be obtained from the plains and watersheds, a great deal from the slopes and valleys.

And as valleys are scooped out and hills and mountains lowered, at each stage the work presents some different feature and touches some different formation. The Secondary and Tertiary rocks, that once overlay the Silurian, have all in turn disappeared, under the mighty influence of this imperceptible but all-powerful force. And just as one formation is pared off, and one surface-soil merges into or is changed to another and a different one, with all the attendant variations of climate and physical surroundings which must follow, the plant-life on every soil is undergoing its like changes and modifications. The grasses or plants are constantly varying, or changing altogether—old ones dying out here and reappearing there, perhaps, and new ones coming up in other places. Pastures, then, are very far from being permanent things. Some are improving and others are degenerating, just as circumstances herein briefly touched upon affect them. The moss-covered soil, with its decayed grasses, may even now be vivifying for a period of more enduring fertility. But on the hills the work of waste is much more active than that of renovation, and the less the soils are capable of yielding the less they can afford the waste that goes on in the poorest of hill pastures, when it is grazed and fed without restitution of its materials being made by the farmer.

CHAPTER XX.

HEATHER BURNING.

On mountain grazings a very large proportion of the land produces heather, and heather only. On this ground black-faced sheep manage somehow or other to pick up a wonderfully good living, which proves that heather must possess a certain amount of nutriment suitable for the food of stock. At the same time a great number of acres of it are required to maintain a sheep. From three to six acres per sheep is about the regular stocking on such land. But there are different qualities of heather, as well as different qualities of soils, and its merits, like that of other crops, depend on the soil upon which it grows. Moss heather, or that which grows upon a mossy soil, is considered quite as valuable as any pasture composed solely of grasses. In fact, some shepherds declare that young heather of this kind is as fattening for sheep as the best of clover. Its value, however, greatly depends on its age. When heather is allowed to get old it becomes very woody, and unfit for sheep to eat. It is necessary to burn it every few years, and the best way to manage it so as to keep it up to its highest standard for grazing purposes is to burn it in regular rotation. It is of no value for a year or two after being burned, so that it is not expedient to burn too much in any one year. A keeper on one of the most extensive estates in Scotland once related to us how he secured his position. The noble proprietor arranged for an interview, and the first and last question he asked was, " How would you burn heather?" " In patches, your Grace," was the answer; to which the Duke replied, " You're the very man I want."

No fixed rule can be laid down that will suit every moor, nor will the same rule suit all the ground on even one moor; it all depends on the nature of the ground at what interval it should be burnt. Where the hill is a dry peaty soil, with pure heather and little or no grass among it, if the heather is burnt early in the spring, that same summer there will be a sprout from the roots, the next year a green carpet, and the second year flowers and food for sheep. About the fourth or fifth year it will shelter birds. This is on ground most favourable for heather. On other lands, after it is burnt it will take three or four years before the sheep can feed on it, and probably eight or nine before it will shelter birds. The grouse have always to be taken into account in this question, more so than the sheep. Again, if there is a good deal of grass growing among heather before it is burned, the grass often comes away sooner and thicker than the heather, and seems to choke it; it depends a good deal on the season immediately following the burning whether the crop will be heather or grass. Different soils require to be differently treated. Without knowing the nature of the ground it is impossible to lay down a rule for burning, just as it is impossible to plant and grow trees without knowing the nature of the soil. It is probable, however, that in no case can justice be done to the grazing where the heather is not burned once in every ten years at the least. Where it is only burned once in say twenty years, the farmer has lost at least ten years' use of his ground.

If heather is burnt in the autumn, and a portion of the same ground burnt in the following spring, the spring-burnt heather will come away the soonest. In autumn the ground is drier, and the fire strikes deeper than in spring; this, and the ground lying bare to the frost all the winter, seems to paralyse the roots for a season. It is the law in Scotland that no heather can be burnt after the 10th of April, that is when the birds begin to nest. The heather on high ground should not be so much burnt as on the low ground. There are moors in this country in the proprietor's own hands, where few or no sheep are allowed, and there are adjoining moors regularly let for sheep as well as grouse; but, so far as regards sport and number of birds killed, there seems very little difference between the sheep-bearing and non-sheep-bearing moors. There are many moors in the north and east of Scotland where it is almost all heather, with little or no grass, while the moors in the west and south of Scotland have a great extent of grass. The north and east country moors, where heather is abundant, will, acre for acre, bear far more burning than the grassy moors of the west and south.

Heather burning has always been considered a "game question," and proprietors invariably retain the power of burning in their own hands. But there is no reason why this should be so, as what is good for the sheep is also best for the grouse, only the landlords have been slow in recognising this fact. Many instances could be given to prove that where the heather is regularly burned and kept under systematic rotation, it is best alike for sheep and grouse. A certain writer says, "I have seen at least one instance of the good result of an accidental, but very extensive, case of moor-burning. It happened on an estate in Inverness-shire about thirty years ago, and burnt a tremendous extent of rough heather. The number of acres has escaped me; but in the course of a few years thereafter the 'Lundy Beat' was regarded as the best of the whole estate for grouse. The soil being naturally moist and mossy, the heather grew quickly, and nowhere else could so large or so good a bag of fine healthy birds be made as on this ground. All this goes to prove that the interests of sportsmen and graziers are identical. I have always thought so, and have no hesitation in saying that the system of burning one-eighth part of the moors every year in strips, as now practised by a few proprietors, should be universally adopted. One of the great advantages of judicious burning is to draw sheep away from the meadows and shielings below, on which they are so apt to dwell in the early part of the season. Nothing interferes and fags sheep more when weak and emaciated in spring, than toiling through fields of old over-grown heather; and it is astonishing how a series of long burnt strips through these rough lands facilitate their movements up and down the mountains. Without such passages it is impossible to do justice to the stock. In situations I have frequently seen the cotton grass, stool bent, and deer hair of the higher levels, so necessary and beneficial at this season, almost entirely unconsumed for want of proper burning."

CHAPTER XXI.

IMPROVEMENT OF HILL PASTURES.

FROM the naturally rugged character of the hill grazings of Scotland, and the want of a cheap and practicable means of improvement, it is feared they must largely remain in their original unfertile condition. Naturally enough, pastoral farmers, keeping in view the raising of "two blades of grass where one grew before," have at various times ventured to exercise their skill and means in bringing them into a more fertile condition; but it must be admitted that no great success has ever attended their best directed efforts. In many instances of so-called improvement, there is ample evidence of a tremendous sinking of money to no purpose. On the other hand, where rational means of improvement were adopted, a great deal of good has been accomplished, and the results have not only been beneficial to the stock, but highly profitable to those concerned.

Drainage, which is the first principle of all good farming, has proved to be a profitable operation on mountain ranges, and so has the erection of fences, shelters, and the enclosure of meadows for hay. But other improvements in the shape of irrigation, manuring, or breaking up with the plough, have mostly all ended in failure and disappointment. Irrigation, so much in vogue at the beginning of the century, has, like all other experiments begun without attending to local circumstances, been quite laid aside. After the advent of railways throughout the country, liming was largely practised in many hill districts, but that method, too, while having much to recommend it, has unfortunately, in the majority of cases, also been abandoned, owing to the expense proving greater than the benefits derived from it.

For a few years after the application of lime to hill land, it has a most wonderful effect in promoting the health and condition of the sheep, but from the thin and porous nature of most soils, and excessive rainfall, the effects of the lime are of a very temporary character, it being prematurely wasted away. Neither has breaking up with the plough brought about any advantageous improvement. Some astonishing results were at one time published regarding the reclamation of hill pastures, and it was then believed that the whole system of hill farming was about to be revolutionised by this means, but it is doubtful if in any single instance the land has been permanently improved.

Several individuals broke up patches on the hills, which they usually first limed and then seeded down with an improved variety of grasses. The Duke of Sutherland entered into this work of reclamation on an enormous scale. He drained, ploughed, and otherwise improved 7000 acres on his estates in Sutherlandshire. That was only ten or twelve years ago. But like the smaller experimental plots of humble individuals, the whole of this land has reverted to its former condition, and its present condition is even worse than its original state before the supposed improvements were effected. It is seldom that we hear of any one breaking and liming now-a-days—that is, of course, on strictly mountain pastures. Where the land is afterwards fenced and cropped in rotation is quite a different thing.

Whatever means are adopted for the improvement of hill pastures, they require to be in sympathy with the natural conditions of the locality. It is useless to attempt to alter the fertility of the soil, beyond what it is capable of sustaining for some considerable time. Climate and elevation are established forces; and compel us to recognise their laws. It has been proved, many times over, that cultivation cannot be carried on with profit above a certain level, and that has already been reached, and in some cases over-reached, by the present boundary line of arable farming. In future, improvements of hill lands must be conducted with some degree of reason, and not on a haphazard system, as has been too much the rule.

Drainage.—In the practical application of any system of drainage, a reference to the nature of the soil, and the plants which it bears, must form one of the primary considerations. Another matter of consideration ought to be what kind of plants it may be requisite to retain in the soil; and what effect draining will have in retaining and improving certain varieties, and in deteriorating and wearing out others. The position of the soil to be laid dry, the connection it may have with the drier grounds by which it is surrounded, and the rainfall of the district, have also an important bearing upon the question of drainage. In a district where the amount of rainfall is small, and the soil in general easily affected by drought, there is always a greater demand at certain times for the softer plants, the produce of damp land, than there is in tracts where a heavy annual rainfall occurs, and where more drains are required to carry off the water falling within that compass. All rules, however, must be modified to suit the various conditions met with on different soils and in different situations.

A stiff clay soil resting upon a slope or ridge, and producing only the coarser grasses and flying-bent (*agrostis vulgaris*), which are drifted off the land by the dry winds of spring, is not improved by too close draining. The accumulation of vegetable mould is comparatively small on such soils, and, when closely drained with open drains, instead of producing plants in greater abundance and of a finer quality than those formerly predominating, it rather tends to make the herbage coarse and unpalatable, while some of the more succulent plants which it formerly yielded, and which are valuable in the dry months of spring when other food is scarce, often disappear from it altogether. Moderately light drainage is, however, advisable, and will usually have a good effect upon this description of soils.

The case is very different with clay soils lying upon a comparatively level surface, from which the water does not readily flow off. Here the plants are generally of a different character, the vegetation is more abundant, and from falling down annually and decaying on the spot, there is a deep covering of vegetable remains overlying the subsoil of cold clay. This land requires and improves by close draining. The tall spretts and rushes are rendered finer, and in some situations, where the soil can be thoroughly dried, disappear altogether. A different species of plants spring up in their place, and cover the ground with a dense sward of nutritious grasses which afford pasturage of the best description.

There is another kind of bog land with an entirely different subsoil to the clays just mentioned. This is composed of peat moss, overlying a retentive subsoil of a different character. Peat bogs are often very rich in palatable grasses, and although they may in many cases be drained with advantage, there is a danger of destroying some of their most valuable plants, which only flourish under certain conditions. The rush and sedge (*carex*), which are in much request in the early months of the year, are more succulent and palatable upon this soil than upon any other, and care must be taken not to overdry it, else the main support of the sheep may be cut off at a period when no other food can be obtained.

Farms on which there is a considerable extent of moss-land invariably produce the best sheep, and while it may be difficult to artificially promote the growth of such plants as are indigenous to that class of soil, it is easily possible to destroy them by the same means. A great proportion of this land occurs in basins on the mountain tops, and having been formed by the decomposition of aquatic plants, possesses elements of great fertility. Where the surface is green, no description of hill land improves more by draining than loose mossy soils, but the case is widely different when the soil is covered with heather, deer-hair, cotton-grass, and stool-bent. These plants form the staple food of hill stocks in spring. The latter of them have long and succulent roots, which the sheep readily draw out of the wet land; but when once it is laid dry, the roots not only get smaller and drier, but, owing to the moss getting firm from the draining, they cannot be pulled up; and as the best part of the plant is that which is below the surface, it is consequently lost. The draining of solid peat-moss is seldom attended with any advantage, except in some peculiar situations where there may be stagnant water, or where it may be necessary to divert the overflow from damaging the land adjacent, which may be of a different description, kept in a wet condition solely on account of the other. These are the general features to be studied in draining; but experience will teach that they are of endless variety, each requiring an equal amount of consideration.

On hill pastures drains are only useful in removing the excess of surface water, and for that reason it is not desirable to put them at so great a depth as is necessary on arable land. When the water level is maintained near to the surface of the ground, the moisture has the effect of keeping up a more equable temperature to the roots of the grasses, as well as supplying nutriment to the blades. Surface or open drains are, therefore, better adapted for hill pastures than tile or deep-closed drains. A more palpable reason, however, in favour of open drains is, that the expense of construction is trifling compared with tile drainage. Even had the latter been more serviceable, it is doubtful if the benefits derived would justify the cost. In some parts of the country there is a preponderance of hard, tilly soil, which cannot be rendered dry by means of open drains, as they do not reach the flow of water which runs along the top of the impervious subsoil : and on such land there is no doubt that tile draining would have a much more beneficial effect. But the question of expense operates sadly against this method being adopted, and more especially as, where it is most needed, the subject is at the best of a very inferior quality. Naturally poor soils cannot be expensively drained at a profit, as their yield is insufficient to repay the outlay. Where the soil is of a good quality draining may be performed with more satisfactory results, and a surer return for the money invested.

The system on which the drains are laid out requires a good deal of forethought, so as to render the operation as effective and indestructive to the pasture as possible. It is an axiom in tile-draining to give the drains as much declivity as possible, running them straight up and down the hill ; but with open drains a somewhat different method is often necessary. When the drains are of great length, and the soil of a loose, friable nature, the objection to the direct action system of draining is, that the water is apt to dig and run away the bottom. On that plan the drains also impede the progress of the sheep in grazing, more than when they are laid on with a slope. There is no definite rule in this matter, but, as a general principle, they should be cut across the declivity at such an angle as will insure a free but not a rapid passage to the water. Care should always be taken to keep the drains short, to prevent a heavy flow of water wearing them so deep that they become positively dangerous to sheep in low condition, or to young lambs. The number required will depend upon the wetness of the ground ; but by laying them out well fewer will suffice, besides contributing several other advantages so as not to cut up the pasture.

For ordinary purposes, the size of the drain that combines most efficiency in proportion to its cost is of the following dimensions :—22 inches wide at the top, 6 inches wide at the bottom, and 16 inches deep. A less depth is liable to early closing by the growth of vegetable matter in the bottom ; and when deeper, they are, of course, more costly to construct, besides being more dangerous to the lambs in crossing. The main drains may require to be made a few inches wider, but any addition to their depth is only necessary under special needs.

The tools required for cutting open or "sheep" drains as they are called, are few and inexpensive. A large sharp edged spade, the shape of a heart, and a toothed fork or "hawk" for pulling out the sods, is all that is usually required, but it is well to have a pick also for use in stony places.

The cost of draining is estimated at so much per chain, rod, &c., taking into account the nature of the ground to be operated upon. The price usually paid ranges from a 1d. to 2d. for seven yards, but that varies according to the demand for labour. The time which open drains are supposed to keep in efficient repair depends upon the nature of the soil. Thin soils with rocky bottom grow up most quickly, while on clay subsoils their durability is greatest. They require to be gone over about every ten years, but the shepherd should always repair any overflows or stoppages that may occur during the interval. Where much draining has been effected the burns are quickly flooded after a heavy fall of rain, and rendered impassable to the sheep. It is therefore advisable to construct bridges at all the dangerous points, and when the sheep are in the habit of using them daily, they are duly protected in the time of floods.

It would be difficult to calculate all the advantages derived from draining; but taking the pastoral districts in general, it cannot be said that more stock has been kept, yet there can be no doubt that the health and condition of the sheep have been much improved. The changes wrought upon a stock by draining are slow and gradual. It is some time before the character of the herbage is altered; and the earlier and more marked changes in the condition of the stock is their greater freedom from certain diseases, the improved quality of the lambs, and superior wintering of the hoggs. Draining has not perceptibly diminished the prevalence of such diseases as louping ill and braxy, but it has had a salutary effect on many flocks that were formerly much subject to rot.

Enclosures.—There are no means of providing for the support of hill flocks in winter, but what must be actually stored up for them. It is possible by means of draining, liming or manuring, to improve the quality, and increase the quantity of herbage grown, but unless the food be preserved during summer, it will cease to exist before winter. This is, in fact, an illustration of the simple proverb, " We cannot eat our cake and have it too." As soon as winter approaches, vegetation ceases to grow, and however well the flock may have fared in summer, they may starve on the same ground in winter. It is absolutely necessary to enclose a portion of the land for the express purpose of growing winter food. By light stocking and careful herding, the best of the bogs on a hill farm can be mown and converted into hay for use in winter; but, while this plan may be practised on grassy farms, there are others on which it cannot, and the system is by no means a good one. There are other resources besides growing hay that the farmer can draw upon, such as turnips, corn, cake, &c., all of which have been used with advantage in exceptionally severe winters. These foods, however, are expensive, and their use involves a system of management, so hostile to the natural habits of mountain sheep, that they should only be resorted to in extreme cases, and used only as auxiliaries.

Hay is obtainable on almost every farm, if only the means be taken to secure it; and most farmers admit that it is by far away the best winter food for hill sheep. Yet, ready as are the means, and wide as is the area from which this valuable product may be obtained, its cultivation is, nevertheless, by the majority of store farmers, neglected. Were our winters all alike, mild or severe, probably a more even course might be chosen in regulating the supply of provisions; but the difficulty is, that a severe winter is frequently followed by several of a milder character. Farmers then forget themselves, and fail to husband their resources, so that when a bad winter does arrive, they are unprepared for it. After growing a supply of hay, a man may be excused for putting up an extra lot of cattle, on the assumption that the sheep will not require it; but one who neglects to provide it altogether, trusting that the winter will be as good or better than its predecessor, pursues a foolish course, and deserves no sympathy. The poor sheep are to be pitied, but not their owner's ruin. The only safe plan to adopt in this matter is to keep a year's supply of hay always on hand. At the beginning of every winter, there will thus be two years' produce in store, and the winter will be severe indeed if more than that is required. Should two or more severe winters follow each other, of course the position of even the most prudent manager may be far from secure, but that is a risk which has to be run.

Hay grown on enclosed pastures is superior to that cut from among the sheeps' feet. The latter consists only of coarse grasses devoid of nourishment, which have previously been picked over by the sheep in grazing; whereas, the former contains all the finer in abundance, and is as

nutritious as summer pastures. As to the number or extent
of enclosures required, it has first to be decided what
system of culture is best adapted for producing the crop.
Hay cannot be grown on upland meadows year after year
without manure. To overcome this difficulty, an extra
number of enclosures are sometimes made, cutting and
grazing them alternately, but the better plan is to manure
wherever possible. Dry land, liable to be injured by
summer droughts, requires, early in spring, an application of
nitrate of soda to force an early and rapid growth. The
dung derived from cattle kept for consuming any surplus
hay, if applied in the autumn, will assist in maintaining the
fertility of the meadows, but for extirpating moss, and promoting greater luxuriance, all land for hay, whether bog,
meadow, or dry, requires a dressing with lime every twelve
years or so.

A supply of 5000 or 6000 stones of hay may be reckoned
a sufficient quantity to provide annually for a flock of 1000
sheep. Estimating the yield to average about 180 stones
per acre, the extent of land necessary to enclose can be
easily calculated, whether for cutting annually or tri-annu-
ally. A few enclosed fields about a hill farm are always
useful. They are serviceable for maintaining the tups, in
recruiting the leanest of the flock in winter and spring, or for
ewes with twins at lambing-time, besides being handy in any
contingencies that may arise in the course of management.
Taking all these advantages into consideration, together
with the actual benefit to the whole flock, the necessity for
enclosures should be clear to almost any one. There is no
doubt that, by fencing off the best parts of farms, the grazing
on the whole is somewhat deteriorated; but so long as the
sheep are allowed the full benefit of what the enclosures
produce, in either hay or pasture, the aggregate profits will
not be diminished. Enclosures are, in truth, the mainstay
of hill flocks, and cannot be dispensed with upon any

account. It is essential, however, to farm them in a careful manner, keeping the fences in good order, and not grudging them a due share of manure. The fences should in all cases be stone dykes, not wire or paling, as shelter is one of the most needful requisites in such enclosures.

In order to secure a good and abundant crop from these enclosures, it is better to dispense with early pasturage, if it can possibly be avoided. In any case, the stock should not be allowed to remain in them later than the middle of May. Well-manured meadows, from which the stock is early removed, may sometimes be ready for the scythe in July. It is of much consequence to have them in rick before the Lammas rains commence. When cut early, the hay is generally of a better quality, and the aftermath more valuable for autumn grazing.

In making the hay, the usual course of procedure is well understood; but a brief sketch of the operation may nevertheless prove useful. After mowing, which is done either by the scythe or machine, the whole is spread out equally to dry, and turned lightly with rakes, and in the evening put into very small cocks. Next day it is again spread out, turned a second time, and in the evening put into larger cocks. In this form it may safely remain for some days. It is once more turned, but not spread out, and then carried to the driest spot, where it is put up in summer ricks, containing from thirty to forty stones each rick. The rick is secured with two ropes, and it thus remains till it is convenient to put the hay into a winter stack. The dimensions and form of the winter stack are frequently misplaced. A round form does not admit of cutting away portions without loosening the bindings, and when the size is large the hay is liable to heat. Fine meadow hay, early cut, is the worse for heating, although coarser hay may not suffer much injury by that process, and in some cases may even be improved by it.

The stack ought to be of a rectangular form, ten or twelve feet wide, and of any length, placed with one end towards the north, and the other towards the south. A stack of this moderate breadth does not heat, the hay retains its colour and juices, and even the seed and flowers remain sound on the grasses. If only ten feet wide, thirty feet long, and nine feet high when built, reckoning from the ground to the eaves, exclusive of the head, such a stack should contain about five hundred stones of twenty four lbs. each. The stacks ought to stand parallel to each other, at least twelve feet asunder, to allow laden carts to pass between them, and also to admit free circulation of air in every direction. The bottoms ought to be well laid with stones or heather "birns," which will prevent waste of the under part of the stack.

Fences.—In the improvement of hill pastures, fences come next in order of merit to draining. The advantages of a well-fenced farm, whether arable or pasture, are so apparent, that it should be almost unnecessary to recount them. In the first place, fences ensure quietude and gentle herding, which are of the very greatest consequence to the well-doing of the flock. In open hill ranges, where the boundary lines of different farms are but indistinctly visible, the sheep are constantly trespassing across the marches. The only means employed to teach them to which side of the line they belong, is a severe chasing and worrying by the dogs belonging to the shepherds on opposite sides. Treatment of this kind is, of course, not calculated to benefit the animals so abused; but it has also another equally disastrous effect. The sheep being frightened, are afterwards afraid to approach those parts of the grazing, and the consequence is, that the pasture is only half eaten, or altogether unemployed, for some distance on each side of the line. If the land on each of the adjoining farms was of a similar quality, the trespassing spirit in the sheep would be in a measure subdued, but it frequently happens that inferior pasture on one farm lies adjacent to good ground belonging to another. The result is, that the sheep not knowing any better, and no fence to prevent them, are very liable to be overcome with the temptation, and cross over on to forbidden ground, where they remain until satisfied; if caught red-handed in the act of stealing, then woe betide them. Without taking into account the loss which is thus sustained by the sheep, these occasions have often led to serious quarrels, not only betwixt the shepherds, but between the farmers as well, rendering the neighbourship both contentious and disagreeable. Should the friendly relation which ought to exist between neighbours become interrupted, instead of assisting strayed animals back to their own ground, they more frequently direct them further from it, and in this way a great many sheep are hopelessly lost. All these annoyances, however, are avoided when the marches are protected by good fences; and their erection is absolutely imperative, before attempting to carry out any system of successful management.

Where natural barriers exist between the different farms, in the shape of ravine or height, there is then perhaps no need to construct costly fences. But these acquisitions are the exception, rather than the rule, on most estates. It is doubtless the part of the landlord to erect a suitable fence around every farm he owns. Under the ordinary endurance of a lease, the tenant would not be justified in exacting so expensive a permanent improvement, but he certainly ought to bear half the expense incurred in keeping the fences in repair. These matters are best adjusted at the beginning of a new lease, and the man who enters a farm without making his tenancy regarding fences as secure as possible has only himself to blame.

As to the material most suitable for building march fences

that will depend upon locality. Dykes are by far the most
preferable, and where stones can readily be procured for
the purpose, no other material should be used. They afford
valuable shelter to the stock at all times, and although a
certain amount of danger attends them in drifty, wintry
weather, yet the benefits they procure are considerably
greater. It is true they are expensive to construct, but
when the stones are once collected, they remain for all
time, and the only future expense they involve is in keeping the walls in order. Stone dykes are, of course, liable
to fall out of repair, but in the course of a nineteen years'
lease, if in good condition to start with, they will require
very little mending during that time. The cost of erecting
a stone dyke will vary with the facilities for procuring the
material, but leaving that out of the question, the price for
building runs from 2s. 9d. to 3s. 6d. per rod of 5½ yards,
according to the height desired. From 5 to 5½ feet is a
suitable height to build, but it is sometimes also necessary
to run a wire along the top of the dyke, where the sheep
have acquired the habit of "louping," which they readily
do if the walls are allowed to fall out of repair. In erecting
the fence, great care ought to be taken to build upon a
solid foundation, otherwise the wall is apt to incline to a
side and gradually fall down. The coping should be made
close and compact. To bed the stones sufficiently, and to
give each a hold of the other, are likewise matters of importance, the duration of a wall depending entirely upon
the attention given to these particulars.

Wire-fencing has largely superseded dyke-building of
late years, and it must be said that the innovation is, in
many respects, a great improvement upon dilapidated dykes
or no fences of any kind. The cheap cost of wire along
with the necessary stakes, places material of this kind in a
much more favourable light than stone. The carriage of it
to distant situations is trifling compared with the other, so

that, upon the whole, the wire will be more generally used. A wire fence, intended as a thorough and complete barrier to sheep, requires six strands at least, with stakes every seventh foot apart. Very many fences on hill ground, however, have only five wires, and where the pressure against them is not very great, this number gives every satisfaction. By using barbed wire for the top strand, a great deal fewer stakes are required, and if the fence is constructed entirely of this wire, which there need be no hesitation in doing, the cost of erection can be reduced nearly one half, as not only will fewer stakes suffice, but fewer strands of wire are required. The average cost of erecting a wire fence is about 9d. per yard, exclusive of carriage of materials. Those having any great amount of fencing to construct cannot do better than take estimates from firms who undertake this kind of work.

What has hitherto been remarked refers exclusively to march fences, but on hill farms it may sometimes be good policy to divide the several hirsels by the same means. On the high-lying farms in the Highlands of Scotland, a great deal of unnecessary herding might be avoided were the pastures properly sub-divided by fences. It would enable the grazier to make better use of the pasture at his disposal, at the different seasons of the year. Instead of having constantly to keep driving the sheep from the low ground in the summer, a fence would stop them at once, and far more effectually, which the sheep would soon recognise, and afterwards settle to feed on the high ground, greatly to their own advantage, while the low ground would be preserved and in good condition for feeding during the winter months. It would be a mistake to attempt to confine the sheep too much on small holdings, but when the extent of the grazings extends beyond what the sheep can travel over with ease every day, the fencing might be of considerable advantage. If possible, give the sheep a moderately wide range. On some farms they are made to travel six or eight miles every day, which is too far, as they have neither time to feed nor to take sufficient rest. Four miles a day is as far as a flock should travel in the course of grazing.

Sooner or later, we believe, hill lands will be sub-divided and interiorly fenced, as the readiest, cheapest, and most effectual means of improving the pastures. Every one who is acquainted with the condition of hill pastures must have been struck with the immense quantity of coarse dead herbage which is never fed down by the sheep; and is worse than so much waste, because it harbours insects and other vermin, and is not without reason suspected to be a fertile cause of sheep disease. By feeding the entire ground quite bare once a year there would be an annual spring of young and succulent herbage which the sheep could relish; the result, indeed, would be very much the same as if the land had been well sweetened by liming. This cannot, of course, be done without fencing and heavy stocking, at one season or another, of the various subdivisions of the pasture. By doing so, however, the pastures would not only be rendered healthier for sheep, but they would carry far more stock than they now do, and thus disprove the commonly accepted opinion that the majority of hill pastures are already too heavily stocked. It is preposterous to talk of hill lands being overstocked at present. There may be too many sheep on them for the system of grazing pursued; but the mismanagement of the grass and the capabilities of the pastures are two different things altogether. Some rough pasture would have to be reserved for winter, if the ground was fed bare, as we propose; but it would be grass of the season's growth, one or more divisions being summer "hained" for this purpose. Then, and not till then, will hill lands be farmed with full profit, and the excessive death-rate amongst hill sheep be diminished.

Plantations.—Among the many improvements which have been attempted upon mountain grazings, these are of more real benefit to the flocks than any other kind of shelter, and yet none have been more sadly neglected. With two or three exceptions, the proprietors of Scotland have cut down far more trees than they have ever planted. This neglect chiefly arises from the circumstance, that those who plant timber rarely live to cut it. It is, however, matter for regret, that at a time when we are likely to be deprived of all foreign supplies, so little attention is bestowed in planting trees for future consumption, and, at the same time depriving the nation of a product which would be of the very greatest benefit to mountain farming. The Duke of Athole and the Earl of Seafield have set a good example to others in the same position, by adding thousands of acres to the woods on their estates, and there is no fear but they or their successors will one day be amply rewarded for the outlay involved. The Duke of Sutherland has also been doing good work of this kind on his farms at Shiness, and so has Mr. G. G. Mackay of Glenloy, Lochaber, who has planted large areas at both extremities of his property, for the purpose of affording shelter to sheep. Moreover, this gentleman is persuaded that, from every point of view, there is no kind of improvement on an estate at all to be compared with planting. It is, at the same time, "the most beautifying, most beneficial, and most profitable."

When a plantation of timber has to be formed, the first step necessary is to fence the ground about to be planted, so that cattle of all kinds may be kept from making inroads. If wet or boggy, open drains ought to be dug throughout the area, and the surface made perfectly dry. These operations being performed, the planting may proceed, in executing which, great care should be taken to make the pits of a proper size, and in filling them up, that the best earth be returned nearest to the roots of the young trees. A mixture of timber in the same plantation is always advantageous, and thick planting is eligible for the purpose of affording shelter.

As the plantation gets forward, great care should be paid to thinning and pruning the trees, removing those first that are either sickly or debilitated; and in this way, and by exercising constant attention in the management, timber trees will advance with greater rapidity than when neglected or overlooked. It has been aptly said, that much expense is often incurred in planting trees, which is afterwards lost by neglecting to train them up. Trees, indeed, are in too many cases merely put into the earth, and then left to grow or die without further attention; whereas with them, as with all other plants, the fostering hand of man is indispensably called for in every stage of growth, otherwise they will rarely arrive at perfection, or make that return to the owner which may be reasonably expected, when the several processes of planting and thinning are duly exercised.

Stells and Kebhouses.—Every hill farm, to be well equipped, and in an efficient state for the safety and protection of the stock it maintains, must be provided with a suitable number of stells and kebhouses. They are as necessary on a mountain sheep-farm as the steading is for an arable one. They ought to be considered as permanent fixtures belonging to the holdings, and erected at the expense of the proprietor. Those who have not a sufficient number on their farms must blame themselves for not seeing to it before signing their leases. The expense of maintaining them should be equally borne by landlord and tenant; the former would then be entitled to a higher rent, and the latter would reap his reward from his stock being kept in much better condition, and being saved many deaths during both winter and the lambing season.

There are different kinds of stells, however. Plantation

stells are formed of tall, close-growing, evergreen trees (such as spruce firs), which, when grown, make a splendid shelter of themselves, as they catch the snow on the outside from whatever quarter it blows. A cart entrance is left open on two sides (generally on the north and south) for driving in cattle or sheep, and carting hay, &c. In the centre of this plantation the "stell" is formed, and consists of a wall about eight feet high (round or square), with gates at the two entrances. From this wall a roof slopes inwards, twelve to twenty feet wide, supported on pillars; an open space being left in the centre. Under this roof racks are fitted for hay, while outside, close to the gates, stores—hay and straw—for food and bedding are accumulated before winter commences. On large farms more than one such "stell" would be required, into which the sheep are driven during storms; the amount of heat and comfort the poor animals obtain under these roofs is wonderful when they huddle together on the "bieldy" side. Of course the size of these shelters must be regulated by the number of sheep on the farm. Considerable care is required to select convenient and suitable sites near to the shepherd's house, or in sheltered situations that are not liable to be blown up with snow. In situations where plantations are awanting (some people object to trees altogether, considering them more dangerous to the sheep in drifty weather) the "stells" are constructed of dry stone walls; where stone cannot be had, the turf or paring of sheep-drains will be found a good substitute. The form of the stell is circular, and for an ordinary cut of sheep, numbering from eight to ten scores, about fifteen yards in diameter, having a small entrance gate and wing dyke, of say three roods long, running at a tangent from the circle from one of the scuntions of the gateway. An ordinary dyke, of say five feet high, will generally be found sufficient. The site of such a stell is a special point. It ought to be thoroughly dry, and the sward such as will not be easily trodden into mire by the sheep's feet. But the site must be chosen in view of all the purposes for which it is erected, and with special reference to the prevailing directions and courses of blasts and storms. There are few experienced shepherds but could tell the most suitable spots on their daily beat for planting such conveniences.

Within the stell, and near the gateway (one of its walls forming part of the stell dyke), there ought to be a kebhouse —a small oblong house built of stone and lime, say fifteen feet long by twelve feet wide within the walls. The walls may be eighteen inches in thickness and six feet in height. A door at one end would give entrance, and all the light required. The roof, to be at once efficient and cheap, may be of corrugated iron sheets, slightly curved to carry off the rain, resting on built-in wall plates, and supported in the stretch across by a few light rafters laid longitudinally from gable to gable, into which they would be built to secure the adhesion of the roof to the walls. The whole furnishing required in the kebhouse would be a few portable small flakes to partition off such space as might be required to isolate an unfortunate ewe and lamb. A space in front of the stell, of say half an acre, ought to be enclosed by an inexpensive fence. The whole cost of such an erection as I have described, exclusive of the inside divisions of the small house, the fence of the enclosure, and the cartage of materials, ought not to exceed, say £15.

As to the uses of such erections, and their advantages to the stock on exposed hills, first may be named the perfect means of shelter they afford from severe blasts and storms. The shepherd, seeing such a storm threatening and imminent at night, would drive his sheep to the shelter, putting inside as many as it will accommodate, leaving the remainder by the lee side of the stell. No doubt a severe snow-drift might accumulate wreaths at the dyke back, from which some of the sheep might have to be assisted out in the morning; but the number of the sheep and the space along the sheltered side of the stell both being limited, the fatal results, and even the labour of searching, would be reduced to a minimum. Those inside the stell would, by their motions, trample down the snow as it fell, and thus be quite safe of being smothered. The stell, however, would serve a great many useful purposes besides acting as storm shelters. Indeed, all the necessities which arise almost daily, where the housing of individual sheep or lambs is required, such as dressing diseased feet, attacks of maggots, as well as for cuttings, speaning of lambs, for washings, &c.—for all such purposes the stells would offer the readiest and best accommodation for their respective cuts. The kebhouse, however, would be mainly beneficial in lambing time. It is no exaggeration to say that scores of lambs are lost from every hirsel annually for want of efficient kebhouse accommodation. " Conceive of anything more sad and trying" (says a good authority) "to a faithful shepherd's heart than to meet in with, in the evening, a 'forfoughen skigh' gimmer with her newly-dropped feckless lamb. The mother, having passed through such an ordeal, is quite callous to the claims upon her natural affections, and, were she able, she would much rather bolt away and outrun the dogs to the top of the hill. The lamb is all but in a state of coma, entering upon 'the sleep that knows no waking.' What is he to do? He might leave the pair to the chapter of accidents and the tender mercies of the cold night. To do so would be to provide for finding a dead lamb in the morning, and the mother as far from her offspring as she can get. He might put the lamb in his plaid-neuk, and give it the warmest corner next his own warm heart, and by the help of the dogs endeavour to drive the exhausted mother home; but even if he succeed, after hours of toil, the result must be the thorough prostration of the ewe, and with the best treatment the flow of milk is restrained, and it will be days and days before she will look at her own or any lamb."

This is no fancy picture; it has often to be gone through, and the loss resulting is not easy to estimate. The keb-house at hand—the much-needed maternity hospital—the lamb, with the mouthful of warm milk from the flask carried in the bosom for the purpose, will wake up all right in a little in the sheltered parrack of the kebhouse; and the mother, in the same snug quarters, catching the first feeble murmurings of the helpless creature, the maternal instinct will by degrees assert itself, and in the morning everything will be so adjusted that mother and offspring will be ready to meet and welcome the earliest sunbeams on the braken brae. During the autumn and winter the kebhouse would make a useful hay-shed, according to its capacity.

Our hills are not without stells of a certain kind, but very many more of the right kind, and planted on the proper spots, are urgently needed, and with the essential adjunct of the kebhouse. Every farmer knows how much better animals thrive with heat and shelter than being exposed, shivering in the cold; and it is astonishing to see how soon both sheep and cattle that are accustomed to shelters naturally go to them on the appearance of storms. Farmers should take up this matter and discuss it at their meetings; but if the Highland Society would embrace it in their valuable operations, and offer prizes for the best constructed and most complete shelters—say, erected in certain parishes or counties within a given time—great good might follow, not only to the landlords and tenants, but to the public at large, from the increased production of human food. Where grown-up plantations already exist, stells could be built in "bieldy" places at very small expense.

Mole Catching.—Hill farmers as a rule wage incessant war against the mole, but it is very questionable if it is advisable to do so. The animal is either a foe or a friend of the farmer, and it is strange that so trifling a matter has not been settled before now. Many people hold that the mole does far more good than harm, with whom we concur, and a great deal of evidence can be produced in favour of this assertion. A correspondent, writing to a weekly journal, gives it as his opinion that this animal is of a decided advantage, at least on grass land. He compared meadows of equal fertility, and, supposing the moles were destroyed in the one, and left undisturbed in the other, what do we find at the end of five or six years? Why, he says, where the moles are killed the grass becomes coarser each year, many of the finer kinds of herbage becoming extinct ; in the other meadow just the reverse is the case ; and wherefore is this? Because the wire-worm and the earth-worm, besides other insects on which the mole feeds, destroy the finer kinds of grasses ; also because these grasses miss the surface drainage the mole effects by its little runs. As to cornfields, he considers that the mole does good service in destroying insects, which, if not interfered with, will destroy the plant altogether.

There are few subjects nowadays without their specialists or authorities, who probe every point to the bottom ; but regarding the effects of the mole, it must be allowed that we are as yet without exhaustive information. Its nature and habits have been fully written by eminent naturalists, but none of them speak of the animal from an agricultural point of view. There has, in fact, been only one small pamphlet ever published in the English language treating solely on the mole, and that is written by a mole-catcher. As a guide to destroying moles, this book gives a few useful practical hints ; but of course, the writer being an interested party, he carefully discusses his subject from one side of view. Even then he entirely fails to bring forward a single fact to prove that the mole is a pest. Against this eccentric mole-catcher, however, we may place the opinions of a distinguished entomologist, Mr. Charles Whitehead, who recommends the protection of moles as a remedy for insects injurious to farm crops.

Another entomologist, the late Mr. Edward Jones, writing on the destruction of the cockchaffer grub, says :—"These grubs inhabit sandy and light loamy soils, and lie about two to six inches deep, and may be found in spring by paring off the sods. My place was much infected by brown beetles, but about twelve years ago, some labourers, removing a bank of earth, exposed a bed of grubs several paces in length. Many of them were scattered among the fallen soil ; and one of the men proposed to strip the surface of the bank, which being done, the grubs were seen lying in irregular channels, as if the parent insects had dropped the eggs moving in various directions. The same man informed me that they were the favourite food of moles, and he desired me to observe an end of the bank, not stripped, being covered with mole-hills, for there no beetle grubs were to be found. When opened, his remark proved true, the moles had traced all the labyrinths of the grubs. I took the hint for the preservation of my foliage, and I have ever since protected the race of moles. The brown beetles gradually decreased, and are now rarely seen. I have always found that in hay and pasture grounds, the moles, as soon as the grass is high enough to cover them, run upon the surface, where they find their food in the numerous caterpillars and insects, which in the early part of the summer, crawl out of the earth, and they continue above ground till harvest."

A writer in the *Journal* of the Royal Agricultural Society maintains that moles are probably as useful as crows, and as a witness summons an English clergyman, the Rev. George Wilkins, of Wix, a distinguished agriculturalist, who attributes no small share of his success to his protecting moles.

So far from seeking to exterminate them, Mr. Wilkins buys them from all quarters, and introduces them into his fields, so thoroughly is he persuaded of their usefulness in keeping down wire-worms and other subterraneous pests of the farm. Like our neighbours, he says, "We used to pay the mole-catcher so much a year, but after learning about the doings at Wix, we told the astonished trapper that no more of our coin should he have, unless in the form of payment for living moles! Fifteen years have passed away, and we have never regretted the humane decision. A neighbouring farmer, who laughs at our love of novelties, and denounces our similitude to that great villain, Catiline, who was addicted to the new 'cupidu rerum novarum' is so far a convert that he has reduced the mole-catcher's salary one half. We confess that one summer, our faith in 'the little gentleman in black' was sorely tried. The moles took possession of the garden in such numbers that the vegetable plots were completely undermined. Wife, daughters, and gardeners of the old school were all clamorous for the trap, and a plentiful lack of carrots and onions was predicted as the inevitable result of our 'strange obstinacy' in persisting to see what would be the end of letting the moles have their own way. The doomed crops were excellent. There is a fact which we hope will induce farmers to think twice before killing such friends as we think moles should be deemed."

In his day there was no keener observer of nature than James Hogg, "the Ettrick Shepherd," and he did not fail to take notice of the mole. He writes :—"Duke Henry of Buccleuch was the first to introduce mole-catching in Scotland. He let the moleing of his vast domains to a company of Englishmen, who undertook to clear his land of moles, and keep it clean for a stated period. But what was the consequence? Why, no more than this—that on all the farms that were most overrun with moles, it has reduced the stock at least one-sixth, in some instances one-fifth; and

N

not only that, but it has introduced two exterminating diseases, the pining (anæmia) and the footrot, which in some
seasons have nearly annihilated the stocks on these farms, as
well as the substance of the men who possess them. On
the very hills where the moles had most prevailed, there did
the pining first make its appearance. Our beautiful evergreen
gairs, which were literally covered with molehills every
summer, on which the ewes lay, and the lambs sported, and
on which the grass was as dark green and as fine and finer
than any of the daised fields of Lothian. Alas! where are
they now? All vanished and become the coarsest of the
soil, and thus the most beautiful feature of the pastoral
country is annihilated. If one hundred men and horses
were employed on a common-sized pasture farm, say from
1500 to 2000 acres, in raising and driving manure for a top
dressing of that farm, they would not do it so effectually, so
neatly, so equally as the natural number of moles on the
farm would do for themselves." It may be doubted whether
the shepherd and poet was correct in attributing the disease
anæmia or pining to the want of the moles; but I can con-
fidently second his assertion, that the molehills are a preventive for footrot in sheep.

In referring to what Hogg wrote about the pasture on
the hills becoming coarser after the moles had been killed,
another well-known farmer, Mr. Laidlaw, Bowerhope, asserts
that the consequences were precisely as represented. "The
green gairs disappeared, the soft succulent plants were found
to languish and die, and their name became almost extinct
in the catalogue of plants of the choicest pastures. These
frequently become coarse, harsh, and unpalatable. In place
of the mountain daisy, the sweet-scented vernal grass, the
healthy sheep's fescue, the rich native clovers, the aromatic
yarrow, the spreading rib-grass, which, with their kindred
plants, delight the sight, a quite different and inferior set of
plants frequently took possession of the soil, such as moss

and lichens, tufty hair-grass, bent, and the like. Where sheep-lands have been open-drained, the moles are apt to injure it by working immediately below the drains; their runs often lead out to them, and carry off the water on the land below, thus more or less defeating the very end for which the drains were made. But the injury done in this way must be quite insignificant; and it may be disputed whether it amounts to an injury at all or not. If the lower and upper parts of the ground are equally well drained, any water which the mole may divert over the pasture, from the upper drains, will be drawn off by the drains below; and it is only when water is in a stagnant condition that it becomes injurious to vegetation. No injury is inflicted in this respect.

On meadow-land it is very essential that no molehill be suffered to exist. Wherever their removal is not attended to, the operation of mowing is rendered difficult, and a circle of uncut grass is left around the hills. A good many farmers, however, neglect to have the hillocks spread out, as they ought to, and so they complain that our friend "Talpa" overruns large tracts of the finest land, covering it with hills, and rendering it capable of bearing only foul, unhealthy grass and weeds. It no doubt costs some trouble and expense to keep the hills levelled; but were one quarter of the expense which is annually devoted to catching moles employed in spreading their hills, a very different verdict would be given. When this operation is attended to twice a-year—in spring, when the grass is just beginning to grow, and again in autumn—the molehills do not injure the meadows, but rather benefit them, because by this means fresh earth is brought to the surface and distributed over the ground, which is exceedingly beneficial to the grass. The same remarks apply to hill pastures, and to grass lands in general.

Then as to what is averred about the mole eating and destroying the roots of crops, the animal is not at all her-

bivorous, and it will die if fed only with the roots of culti-
vated plants. In fact, the mole never touches a vegetable,
and is of strictly insectivorous habits, as the order of mam-
mals to which it belongs would imply. Its food consists
chiefly of worms, slugs, beetles, wireworms, and other larvæ
or grub which may fall into its burrows or runs. In destroying these the mole does good service to the farmer, and far
more than repays any apparent injury it does him. The
formidable hordes of grub and worms, which burrow a few
inches deep in the soil, and prey on farm crops, would be
immensely greater were they not kept in check by the mole.
The little quadruped is constantly digging in pursuit of
them, and the presence of moles in a field is a very good
sign that they are needed there ; for unless there were abundance of grub and insect life in the soil of the field, the mole
would not frequent it. When he finds these getting scarce
he will shift his hunting-grounds, and go where he is more
required.

Salt for Hill Pastures.—It is believed by many that
the application of salt to hill pastures has a most beneficial
effect in preventing many of the disorders to which sheep
grazing thereon are liable. Mr. H. Thompson, an eminent
veterinary authority in the north of England, recently stated
at a meeting of the Scottish Metropolitan Veterinary Association, that he had watched the effects of something like five
thousand tons of salt applied to land, and he found that
practically it had the prophylactic power of preventing such
diseases as red water, fluke, louping-ill, hoose in calves, &c.,
which were more or less due to micro-organisms in the
blood, as Pasteur and others affirm ; and the views of these
authorities on the germ theory are confirmed by the results
obtained from the application of salt to land as a preventive of these diseases. Mr. Thompson's district of observation embraced a seaboard of over sixteen miles on the banks

of the Solway, where sheep and cattle were kept and did well on short and scanty herbage without showing the least symptom of parasitic or blood diseases. In fact, so well was it known among the sheep breeders in the Fell districts, that when they brought their flocks down from the hill to winter, they liked to get them as near to the sea as possible, their losses being so few compared to what they suffered from sickness braxy when the stock was folded on inland pastures.

Microscopic experiments have brought to light the fact, that these diseases are due to micro-organisms in the blood. But before disease can be established, the fluids and solids of the body of the animal must be in a certain condition more or less favourable to the propagation and development of these organisms. The moment the vigour of the vital energy is relaxed or lowered from any cause whatever, the spores of these little organisms spring into life, multiply, and flourish. On examining the analysis of the ash of healthy human blood, it is found to contain nearly fifty-five per cent. of common salt. Now, if it is necessary that the ash of blood requires fifty-five per cent. of chloride of sodium for the establishment of vigour and tone to the nervous system and healthy functions of the body, it stands to reason that a want of it must lower the tone and relax the functions of the vital powers, and disease must be the result.

The application of salt to land destroys diseased or coarse rough grasses wherein may be lurking the germs of insect pests; the grasses die, and in time manure the ground, and fine succulent herbage springs up. Secondly, salt destroys snails and small mollusca. Thirdly, the herbage, through the action of the salt, is of better quality; the animals eat it with more relish, and get into their systems certain elements that are necessary for the manufacturing of healthy blood, which acts upon the nerve-centres, giving the system full tone and energy, and thereby warding off disease.

Fourthly, any excess of salt which may be washed by the rains from the land to the streams, thence to the rivers, will act as a preventive of the fungoid disease affecting the skin of the salmon.

Mr. Thompson, about the commencement of November 1884, was asked by the under-steward of Lord Leconfield of Cockermouth Castle as to the advisability of applying salt to some old-laid grazing lands. For several years they lost about ten per cent. of their sheep, which are brought down from Skiddaw about the beginning of November for wintering, and are returned again in April. That year about 470 sheep were brought down, and in three weeks four died. Towards the end of November fourteen tons of salt was applied to forty-three acres, and although the sheep had also the run of sixty-eight acres unsalted, they were mostly found on the salted portion, and very rarely on the other. Afterwards they were removed to another part of the farm—very old-laid lea, unsalted—and during the fortnight five sheep have died; but while on the salted land, from the beginning of November till the end of January, only one succumbed. In the course of his remarks, Mr. Thompson read a letter from Mr. Thos. Horn, Norfolk, who commenced farming near Carlisle some thirty years ago, and who suffered great losses among his flocks from the disease called " cripples," which is analagous to rheumatism. He applied salt to the land, and the disease totally disappeared, and he now attributed the healthy condition of his stock to the free use of salt.

Plenty of other evidence could be given to prove that salt has a salutary effect upon the health of sheep, but the foregoing experience, if not absolutely conclusive, contains at least a great deal of very evident testimony bearing upon the question generally, which farmers might do well to study very carefully.

CHAPTER XXII.

RENTING AND STOCKING A SHEEP FARM.

IT falls to the lot of most farmers to undertake the duty once, if not oftener, of choosing a farm upon which to settle and employ their skill and capital as a means of livelihood. If a farm of the right sort is selected, the farmer's life may be one of pleasure and contentment; but when the investment proves a grinding-mill for his probably hard-earned capital, the path he treads will be very different, leading, perhaps, to financial ruin and the wreck of his future career. Either of these results are possible with men of equal ability as managers, the fatal step being the false estimate made on starting. All this has been seen and felt, and many can testify to the truth of it. In the

Choice of a Farm, the first point to be settled is the size or extent of the holding. The amount of capital ought to regulate this. But the evil is that few men will work on a scale suited to their means. The custom is almost universal for the man who has the means to do justice to a farm carrying a stock of say 1000 sheep to try to take one with 1500; and to accomplish this he will often go the length of borrowing capital to pay for the stock on entry. Now, neither sheep nor any other kind of farming is sufficiently lucrative to admit of the success of this policy. After the rent is paid, and the necessary expenses of working and management deducted from the receipts, the interest

on the borrowed money is difficult to meet. It is true there
are many instances of men taking farms with little else than
borrowed capital, and succeeding in them too; but the
principle is a ricketty one, and has swamped far more than
it has ever floated.

A sheep farm ought to be of such a size as will admit of
at least one man's full employment. After that it may
carry any number of sheep desirable by the same rule. If
the stock is larger than one shepherd can look after, or less
than sufficient to employ two men fully, then there is always
a difficulty in keeping expenses within proper limits. It
may be said that the size of the farm is already fixed, and
such is the case; but in choosing a farm, the possible
economy of labour is of paramount importance, and requires
to be carefully studied. Mountain farms are of all sizes,
and suitable to all purses, carrying from 500 up to 10,000
sheep. No one, therefore, in want of a farm need be at a
loss to find one suitable in size to the capital he can employ in it.

The next point to consider is the situation of the farm.
Distance from market has only a small influence upon the
value of a sheep farm, yet it is worth considering. What
has more to be noted in regard to situation is the altitude
and exposure. These are points of great moment. A farm
in a low, sheltered situation, is worth a great deal more than
one that is high-lying and destitute of natural amenities.
And there is even a considerable difference between the
dark and sunny side of a hill, as regards the health and
quality of the sheep it grazes. Then the nature and variety
of soils and grasses have to be considered; also whether
improvements are needed in the way of draining and fencing, and the condition or state of repair of all existing drains,
fences, and buildings, the farm house, shepherd's house,
&c., included.

Physical conditions are the great starting-point in sheep-farming. In the choice of a locality, then, we have to be guided by geographical position and altitude, both of which materially affect temperature, rainfall, and pasture; and pasture is further affected by the nature of the soil, which in turn varies according to its origin and formation.

The number of sheep which any particular grazing will maintain cannot be estimated merely by observation or examination of the pasture. In this matter most men are guided by the experience of those in occupation of the farm. The best hill grazings do not quite carry a sheep to the acre; while the majority require two or three acres to a sheep, and some a great deal more than that even. By examining the condition of the stock, a practical man may decide for himself whether the ground is capable of maintaining a greater or smaller number. After that point has been determined, and taking into account the quality of the sheep likely to be produced, the rent the farm will afford can be calculated.

Cost of Stocking.—At present prices a blackfaced breeding stock of 1500, at the term of Martinmas, consisting of 315 hoggs, and 1185 ewes and gimmers, would, on a Whitsunday valuation, represent a capital of about £2680, 4s. Assuming the death-rate between Martinmas and Whitsunday to be 13 per cent. of hoggs, and 7 per cent. of ewes and gimmers, the Whitsunday number would be 275 hoggs, and 1120 ewes and gimmers, the latter bringing 80 per cent., or 896 lambs. Thus, the ingoing valuation would probably be—

BLACKFACED STOCK.

		Whitsunday Valuation.		
899	Ewes and lambs, at 41s. 6d.	£1859	4	0
224	Eild ewes and gimmers, at 30s.	336	0	0
275	Ewe hoggs, at 28s.	385	0	0
25	Tups, at £4	100	0	0
1410	Total	£2680	4	0

This puts the cost of stocking at the rate of 35s. 8½d. per sheep.

Returns.—The annual sales from such a stock as the above, after retaining 315 top ewe lambs to keep up the flock, might be as follows :—

300	Top wether lambs, at 13s.	. . .	£195	0	0
99	Second wether lambs, at 7s. 6d. . .		37	2	6
90	Mid ewe lambs, at 13s. . . .		58	10	0
91	Small lambs, at 5s.		24	15	0
195	Draft ewes, at 21s.		204	15	0
15	Shott ewes, at 15s.		11	5	0
5640	lbs. of wool (4 lbs. per fleece), at 7d. per lb.		164	10	0
105	Skins, at 3s.		15	15	0
		Total Returns .	711	12	6

Expenses.—The annual expenses in connection with the same flock would probably be :—

Herding, 10 per cent. of sales	. . .	£71	2	6
Marketing, 2½ per cent. of sales	. . .	17	12	6
Cleaning drains		5	0	0
Dipping, keeling, &c.		12	10	0
Wintering tups		15	0	0
Hoggs wintered away from home, say 270, at 6s. 8d.		90	0	0
Taxes		15	0	0
Repairs on houses and fences . . .		5	0	0
Interest on tenant's capital, and for management, at 8 per cent.		215	0	0
Total Expenses	£446	5	0
Balance remaining for Rent .	265	7	6

Rent per Sheep.—This gives a rent of about 3s. 6d. per sheep, on 1500, which is pretty near the average presently paid for grazings where the hoggs cannot be wintered at home. On the best Lanarkshire blackfaced sheep farms, however, the rents are quite double this figure, some of them a good deal more than that even, and 10s. or 12s. a sheep was not an uncommon rent a few years ago. The winter losses are heavier in the North, and there the present rents may be said to range from 2s. to 3s. per sheep.

An appeal case under a remit from the Court of Session as to the valuation of the sheep stock on the farms of Ardtornish and Duart, heard before the Valuation Committee of the Argyleshire Commissioners in 1885, showed that the rent of sheep farms had then fallen 50 per cent. within ten years, and there has been a still further decline since 1885. In the evidence given, Mr. George Malcolm, factor on the Glengarry estate, rated blackfaces at 2s. 3d. He did not know of 3s. being got in 1885, and would have been glad to take 2s. 3d. for a better farm than the one under dispute. The special causes he assigned for the prevailing depression were—general depression of trade, the enormous importations of meat and wool from abroad, the inflated valuation of sheep stocks to incoming tenants, and the uncertainty as to the future of the land laws.

Mr. Nigel B. Mackenzie said he would rate blackfaced sheep at from 2s. 4d. to 2s. 6d. He stated that Aryhualan, on the Ardgour estate, which was allowed to be the best farm in North Argyle, and formerly rented at £800, was let in 1882 at £550. He attributed the depreciation mainly to the tremendous drop in the price of wool.

It also came out in evidence that a farm in Sutherland, which formerly rented at between 4s. and 5s. per sheep, had recently been let at 1s. 9d. per sheep. Mr. W. E. Oliver valued Cheviots at 3s. rental. On the other hand, Mr. Martin, factor for Poltalloch, thought 4s. was not too high a rent for blackfaced ewes that could be crossed.

The Commissioners, however, decided that 3s. per sheep was at that date a fair rent for the farm.

Both sheep farmers and factors were unanimous that

within the ten years preceding 1885 the depreciation in sheep farms all over the north and west amounted to as much as from 3s. 6d. to 5s. per sheep.

Many of the present grazing leases were entered upon when wool was quite two-fifths above its present value, and that alone must have sufficed to bring about a reduction of grazing rents. But the seriousness of the business is intensely aggravated now that the only other saleable commodities from hill farms, viz., lambs and draft ewes, have declined in value fully a third within the last three years.

Martinmas *v.* **Whitsunday Entry.**—It is very important, when the sheep stock has to be taken by the incoming tenant at valuation, that the entry to the farm should take place at a period of the year when the real value of the stock can be easily ascertained. Many are of opinion that Whitsunday is not the best time. It is too early in the season; the lambs are not saleable; and there are no markets to guide the arbiters in fixing prices; consequently, serious mistakes are sometimes made.

In a paper read at Teviotdale Farmers' Club in February last year, Mr. James Oliver of Thornwood, Hawick, showed that in sixty-seven valuations instanced by him the prices had been 17 per cent., or about 10s. a sheep, too high. No one enjoys greater facilities for arriving at a right estimate of the facts in question than Mr. Oliver, and his figures in the main cannot be disputed. Moreover, at a Whitsunday entry the tenant is obliged to take not only the stock ewes, but the whole cast of the farm for that year, which he has had no hand in producing, and is burdened with the risk and trouble of realising. And this risk is not imaginary; for, as Mr. Oliver proved, those who entered to hill farms in 1884 and 1885 lost in the first six months 20 and 40 per cent. of their capital.

Such risk and loss would be avoided were the entry at

Martinmas, because the cast of the season being all disposed of, the breeding ewes and ewe hoggs would only remain to be taken by the incoming tenant. The season's sales being over, the true market value of the holding stock could be more safely determined, and about one-third less capital would be needed to begin with.

The change advocated will not be easily brought about. Tenants who entered at Whitsunday term are entitled to quit at that term; and as it gives them a larger sale than a Martinmas outgoing they will not forego the advantage unless by special arrangement. The summer sales of lambs, wool, and draft ewes should of course make up the difference between a Martinmas and a Whitsunday valuation; but, as already mentioned, the Whitsunday valuations of the last ten years have been 17 per cent. above what the summer sales warranted.

This is a species of tenant-right which weighs injuriously on the interests of sheep farming generally, and many landlords, we fancy, would be willing to pay the difference on the valuation, as between Whitsunday and Martinmas, if their farms would thereby be freed of the incubus. There is no greater mistake than to suppose that the present system is favourable to the tenants. It requires a third more fixed capital, and the tenant who gets 17 per cent., or any other sum, above market value for his sheep stock on quitting a farm because he paid the same surcharge on entry, is still a heavy loser, though market prices at ingoing and outgoing should correspond exactly. He loses interest on the extra capital invested during all the years he sits in the farm; and, if market prices differ in the meantime, that may be the smallest part of his loss. The sooner, therefore, the proposed change is brought about the better, if it is done with due consideration of vested interests.

CHAPTER XXIII.

VALUATION OF SHEEP STOCKS.

ON hill grazings, where a system of store-farming is pursued, an outgoing tenant is, in most cases, bound by his lease to leave the regular sheep stock of the farm to his successor at valuation price. Even where it has not been stipulated for, this arrangement is frequently adopted in sheep-farming districts. The practice is one that commends itself to all parties concerned, for various reasons. It ensures the outgoing tenant a customer for his stock, and it gives the incoming tenant a stock presumably well-suited to the farm. Much of the success of hill-farming may be said to depend on the last-named condition. If the acclimatised stock of a farm of this description is scattered by sale, it is all but impossible to at once get together again a fresh stock equal to it in all the different ages and proportions; and supposing it could be got together, the chances are that the new stock, as regards size and constitution, would not be half as well adapted to the grazing as the one that was bred and reared on the farm. There is also the risk of a serious loss by bringing a fresh stock on to land that is liable to disease. And it must not be forgotten that, on these wide and open ranges, old sheep, if sent to a strange run, never settle, but continue to stray and wander off it.

It is another matter, however, to say what these advantages are worth in a monetary sense. The new tenant has an undoubted right to as favourable an incoming as possible, and it is for the encouragement of sheep-farming as an industry that it should be so. On the other hand, the outgoing tenant is entitled to full value for the stock he is parting with, but not to more than its value; and if a tenant takes a farm with a fixed stock, which is valued when sheep prices ruled high, he must, all the same, take his chance of the prices that are going when he goes out again. The only rule to go upon is that of honest and capable valuation.

Very grave charges have been made against the honesty of both valuators and those giving over sheep stocks. The latter, in preparation for valuation, are accused of largely increasing the number of sheep a farm will carry, by keeping in aged and infirm sheep; while the former are charged with valuing the stocks exorbitantly high, particularly in the face of falling markets. According to one gentleman who complains under this head, "almost the last thing now required in a valuator is honesty. The object is to secure a partizan. Each nominates the sharpest and stiffest hand at a bargain he can get, and it is not now looked upon as a valuation; it is not the work of two disinterested mutual friends wishing to find out the true value of the sheep, but two diplomatists trying who can outwit the other and buy or sell cheapest or dearest. In the same way the final fight is who shall get the nomination of the oversman."

These charges are far too sweeping. They may have been justified in a few cases, but we have seen more than enough of honest and capable valuation in connection with the transfer of sheep stocks, to justify us in saying that there is only a modicum of truth in them.

In all, or nearly all, leases or agreements which bind a tenant to take to or leave the sheep stock at valuation price, the number of sheep the farm will carry is defined and limited; and, where this is so, the outgoing tenant cannot possibly increase the number of holding stock by retaining aged crock or draft ewes for that purpose. On the other hand, there is little doubt that valuation prices are often 20 or 30 per cent. above open market prices; and it is worth noting that at Meiklehill, Aryshire, last year, the blackfaced stock was valued at £10 per score higher than the outgoing tenant paid for it on entering the farm fifteen years before.

The question, "How far has the prevailing depression reached the pastoral districts of the country?" has never been fairly met and answered. If we take the value of sheep and lambs and wool, as published in the ordinary market returns, and compare them with the prices of, say, ten years ago, we can come to no other conclusion than that the pastoral interest has suffered almost, if not quite, as much as any other interest. In the preceding chapter it has been pointed out that sheep farmers and factors, examined before the Valuation Committee of the Argyleshire Commissioners, were unanimous in the opinion that, within the period mentioned, the depreciation in sheep farms all over the north and west could not be less than 25 to 50 per cent.

But if we take the value of sheep stock handed over from one tenant to another, and valued by farmers who are practical men, it is difficult to imagine that any great depression exists in the pastoral industry of this country. It is too true that the value put upon stock at these Whitsunday deliveries has not fallen in proportion to the market prices. Indeed, one of the causes assigned in evidence before the Argyleshire Commissioners for the depreciation in the value of sheep farms, was the inflated valuation of sheep stocks to incoming tenants. Tenants who wanted to come into the district, complained that they were far more afraid of valuators, because they were unknown, than they were of rent or any other expenses. Loss of sheep was heavy in many districts; rent, they said, they knew beforehand, and it was their own fault if they gave too much; but in cases of valuation—which meant interest on a very large sum of money—they were utterly ignorant. Such a state of things, if persisted in, must cripple sheep farming. Yet a newspaper discussion on this subject, a year or two ago, revealed that the outgoing tenant thought he was not getting one shilling more than he was entitled to, while it was perfectly clear to the incoming tenant that he had to pay a great deal more than he ought to have done.

In saying this much, we make no aspersion on the integrity of the valuators. Doubtless the rates fixed by them have many times been higher than the market prices for hill produce warranted, but in some instances the valuations are considered to have been too low. It is not easy to determine at Whitsunday what prices will be realised for lambs and wool and draft ewes when they come to be sold several months later.

One explanation of the apparent great discrepancy between sale and valuation prices lies in the fact that it is chiefly crock or draft ewes that are sold, the holding stock in most cases being a fixture, as it were, which the incoming tenant is bound to take at valuation prices. Now, stock ewes are fairly worth 5s. to 8s. more than the draft ewes. An acclimatised stock is also worth a few more shillings per head than one strange to the farm. It is well known that strange sheep never do so well on hill ranges as sheep that have been bred on the ground. And crossing, with black-faced ewes, has prevailed so much of late years that young ewes of this breed are not so easily obtainable. We can instance cases where the incoming tenant, not being obliged to take the stock at valuation, has declined to do so; and in the end had to pay exorbitant prices at the sale ring for stock ewes not nearly so good as those he declined, and which he would have found in more respects than one the cheapest ultimately. All things considered, therefore, the incoming tenant is perfectly safe in paying some few shillings per head more for sheep bred on the farm than would be accounted their market value if they were taken off the farm.

It seems to us that if a standard or standards of valuation could be laid down for general adoption, it would put all such transactions on a proper business footing and prevent the evils complained of. As it is, the principles on which the price of stock is fixed are of the vaguest kind, notwithstanding that a vast amount of this description of property changes hands by valuation every year. The elements of value are simple and tangible enough compared with some other subjects of valuation; yet no general standard or method of fixing the values has been adopted in sheep valuation.

Now, it is evident that the value of the stock to the incoming tenant is just what it would be worth to its present owner, if there were not to be any change. The breeding flock ought to be valued prospectively, not with prices based on an estimate of the following summer sales, but as regards their average produce in wool and lambs for the years they will be in the flock, with their ultimate price as drafts, after deducting for annual losses, rent, interest, and all expenses. To do this we require to have a scale price of lambs, draft ewes, wool, rent per sheep, and expenses per sheep; and as flocks differ in quality, &c., there ought to be at least three such scales—top, medium, and low— under one or other of which it would be easy to properly class any particular flock. These scale prices might be fixed annually, in the same way as the fiars prices, and the average of the last seven years' averages should be made the standard, as it is obviously unfair to assign a prospective value based on the last year's sales only.

With prices based on the septennial average scales, and the sheep taken at their average produce in wool and lambs, &c., for the years they will be in the flock, it would be impossible for any of the abuses now complained of to continue. If the flock is assigned its proper scale—top, medium, or low—there could scarcely be any further difficulty; for if over-aged ewes had been kept in the flock their value would be properly discounted. Thus, say the ewes should be drafted at five years old; the two-year-old ewe or gimmer would fall to be four years in the flock, and be estimated to produce four fleeces, four lambs, and the price of the draft ewe, less four rents and four expenses; while the five-year-old ewe would only be one year in the flock, and produce one fleece, one lamb, and the price of the draft ewe, less one year's rent and expenses.

CHAPTER XXIV.

WOOL.

ALTHOUGH the price of wool is low now compared with what it was twenty or twenty-five years ago, when blackfaced clips were worth 1s. 6d. to 2s. per lb., it must ever form a considerable item in the receipts from a hill sheep farm. A heavy fleece of good wool is, therefore, of the utmost importance to the farmer, besides being the best protection to the sheep in winter. The wool prizes given by Mr. Howatson at Perth show this year for the blackfaced sheep, male or female, carrying the fleece best adapted for protecting the animal in a high, exposed, and stormy climate, will do much to encourage improvement in this direction, more especially if the Highland Society continues to offer such prizes annually, and other societies are thereby induced to follow the example. Blackfaced sheep are more noted for length of staple than for fineness of fibre and closeness of pile ; but something more than length of staple is needed to make a heavy fleece, and an open coat, however long and flowing, is ill adapted to protect a sheep from rain and inclement weather. Great improvement has been worked already upon blackfaced wool, but it is far from general; for while in some flocks we find that the staple is 22 to 24 inches long, and the fleeces average from 5 to 6 lbs. in weight—which could not be reached without considerable fineness of fibre and closeness of pile—other flocks fall fully a half short in these respects. The "kempy" wool, or hair, so common in the latter cases, is particularly objectionable, being deficient both in weight and value. The natural coarseness of blackfaced wool unfits it for the manufacture of the finer class of goods, but, as a recent writer has pointed out, " it has fluctuated but little in price for a number of years, the unwashed ranging from 5½d. to 7d. per lb. Immense quantities of it have been exported to America, and this outlet has effectually prevented anything approaching a glut in the market, even at times when higher-priced wools have been scarcely saleable. Glasgow, Leith, and Bradford are the leading centres of sale."

Originally, as history tells us, sheep were principally reared for the sake of their mutton. Considerable value would, no doubt, also be placed on their wool, for sacred history frequently mentions "woollen garments"; but whether the art of wool-manufacture for clothing purposes was known to the antediluvians is matter for dispute. Egyptian priests, who looked to the health of the people, marked the Divine disinclination to use woollen garments by not permitting one to approach the altar with wool upon the person. Though garments of linen could be easily washed, those of wool could not be, and unchanged clothing in a hot country developed skin diseases, so common among the great unwashed. On the other hand, the old religion of India, brought in by pastoral Aryans from colder hills, laid down that the sacrificial thread, the mark of caste, should be of wool. Hipponax spoke, 500 B.C., of the woollens of the Coraxi, and classical writers note the wool of Scythia. The legend of the Golden Fleece seized the imagination of the ancient world, and Philip le Bon, on the occasion of his daughter's marriage in 1429, originated the noble order of that degree, which was so long one of the coveted of European distinctions. Australasia is now consistently called the Land of the Golden Fleece. Three centuries ago Britain was held in high repute for its wool. At that period sheep were raised more on account of their wool than mutton. But the advance of time has wrought great changes both in the character of wool and its industrial purposes.

Growth and Structure of Wool.—The fibre of wool is of a circular form, with a serrated and saw-like edge, which differs in different wools, and even in different parts of the fleece. These serrations, which are exceedingly minute, amount to about the number of 2500 per inch in length of the fibre, and it is to the variations in these that different wools owe their different felting powers. If the wool is of a well-sustained growth, the serrations are full, uniform, and healthy, and the fibre of such wool is stronger and better adapted for manufacturing purposes. The curling power of wool is due to the same cause as its felting power, and is in proportion to the fineness of the thread of the wool itself. The fineness of the wool depends on several circumstances. The size of the thread of the wool depends on the size of the aperture in the skin, or on the pores which are in the skin, and from which the fibre grows. If this aperture is lessened, the wool which passes through it will be lessened, but if enlarged, the fibre of wool is also enlarged. Now, the skin is very susceptible to cold, and by exposure to cold the pores only permit of wool passing through them of a smaller diameter than usual, and so the wool grown is finer and more slender. On the other hand, if sheep are exposed to excessively hot weather, the pores are enlarged, and the thread of wool is enlarged also. Therefore, if sheep are kept in great cold in winter, and are exposed to great heat during summer, some of the wool will be larger in diameter than in other parts of the thread, and it will not be a true or even sample. The value of the sample is lessened by this inequality, as it is more liable to fracture and break in going through the machinery.

Wool is also influenced by the mode of feeding. The

rapidity of growth and the soundness of the wool depend largely on the well-being of the animal. The food required for promoting the growth of wool differs but little from that usually given under any liberal system of feeding. The special requirement is a supply of sulphur, which it usually secures from such green crops and corn as clover, vetches, beans, peas, lentils, &c. The influence which these have on wool has been frequently observed, and we have in this fact an explanation of much of the softness of texture which is then produced. Wool appears to require other materials for growth, but only such as are necessary for the production of flesh and fat. We shall therefore be perfectly safe in promoting the growth of wool—so far as food is concerned —if, in addition to our ordinary supplies of food, we give the animal some variety of the leguminous crops already named.

The finest fibred wools are generally the most elastic. Softness of wool is also of great importance, and so long as it can be maintained without loss of strength, very desirable. The softness of the wool depends much on its fineness, but not always, for the fineness may perhaps be traceable to bad management. Much of the softness of the wool depends on yolk, a secretion thrown off from the skin for the purpose of keeping the wool fine, soft, and supple. It also induces a healthy growth of wool, and, in point of fact, the quality of the wool depends greatly on this secretion— which is really a soap consisting of potash and animal oil. When the growth of wool is rapid, and of a healthy character, there is not only an abundance of yolk in the wool, giving it a soft or greasy feel, but the skin has much the same condition. This is never found upon sheep which are badly fed and in poor condition. Under such circumstances the blood is naturally free from any oily matter, and consequently the roots of the wool cannot get their supply, neither can the skin maintain its soft and greasy condition. A liberal supply of good food is therefore an essential for the production of the best quality of wool. The influence of food does not end here, for a regularity in the supply is almost as important as the quality. Any period of short supplies, or of inferior food, leaves a clear record in the fibre of the wool, producing a harsher and a weaker structure, which is readily distinguishable from the growth produced when the animal is well fed. A lean sheep was never yet known to yield a fleece of the best quality; and where a sheep has been allowed to fall away in condition during the winter, there will be a weakness and irregularity in the fibre of wool corresponding with the period of growth. The microscope reveals the fact that the structure of the fibre is injured by a want of proper nourishment. In well-tended flocks all the variations are reduced to a minimum, because the sheep are suitably fed and sheltered.

Effect on Soils.—In regard to soils, the best wool-growing land is generally that on a sandstone foundation. The wool from it is bright in colour and clean. That grown on limestone is less bright, whilst the volcanic formations produce wool discoloured and stained with red and yellow dust. This is only as regards quality of wool. For weight of fleece, volcanic or limestone soils are the best. Sometimes we find that richness of pasture and excellence of wool go together, where the sheep, fed on rich, deep grass, yield fleeces remarkable for length of staple and quality ; but this is not generally the case. And whatever effect the nature of the pasture may have in altering the quality of the wool, a full supply of food is necessary to give the wool the required length and firmness.

The growing of wool affords another beautiful practical illustration, both of the kind of food which animals require for particular purposes, and of the effect which a peculiar husbandry must slowly produce upon the soil. The quantity as well as the quality of the wool yielded by a single sheep varies much with the breed, the climate, the constitution, the food, and consequently with the soil on which the food is grown.

Wool and hair are distinguished by the large proportion of sulphur they contain. Perfectly clean and dry wool contains about 5 per cent. of sulphur, or every 100 lbs. contains 5 lbs. The number of sheep in Great Britain amounts to 30,000,000, and their yield of wool to 111,000,000 lbs. or about 4 lbs. to the fleece. This quantity of wool contains 5,000,000 lbs. of sulphur, which is, of course, all extracted from the soil. If we suppose this sulphur to exist in, and to be extracted from, the soil in the form of gypsum, then the plants which the sheep lived upon must take out from the soil, to produce the wool alone, 30,000,000 lbs., or 13,000 tons of gypsum. Now, though the proportion of this gypsum lost by any one sheep farm in a year is comparatively small, yet by the long growth of wool on hilly land, to which nothing is ever added, either by art or from natural sources, those grasses must gradually cease to grow in which sulphur most largely abounds, and which favour therefore the growth of wool. In other words, the produce of wool is likely to diminish by lapse of time, where it has for centuries been carried off the land ; and, again, this produce is likely to be increased in amount when such land is dressed with gypsum, or with other manure in which gypsum naturally exists. Large quantities of potash are carried off from the fields in the fleeces of the sheep. This can be recovered by simply washing the wool and evaporating the wash-water to dryness. A fleece weighing 5 lbs. contains about 4 ounces of pure potash, of which nearly 3 are recoverable. This is an explanation of the fact why certain hill lands are known to carry a lighter stock than they did a number of years ago.

Industrial Qualities of Wool.—In order to under-

stand the industrial application of wool, its structure should
be kept well in view.   Each fibre of wool is hollow in the
centre, and covered with a scaly exterior.   It is from this
scaly exterior or roughness on the fibre that wool derives its
most important function, viz., its milling or felting property.   In some specimens the ring-like scales are of great
length when compared with the diameter of the fibre, and
this peculiarity introduces serious difficulties in the way of
the manufacturer, because the solid portion of the fibre
ceases to be elastic and pliable, and easily breaks when subjected to flexure.   In addition to this, all these fibres resist
or are incapable of the felting action, which is so important
a feature in the true wool, and which depends upon the
facility with which the scales of the one fibre interlock into
those of others when in juxtaposition.   These solid fibres
also resist the entrance of all dyeing or colouring matter
into their interior, and will only receive a surface colouring
which is easily removed by washing.   In some cases the
outer continuity of the scales is not accompanied by a
change of internal structure.   Such fibres are usually known
as flat kemps, receiving the dye, but being incapable of felting.
Oval kempy fibres, known among shepherds by the term of
"dead hairs," are most common in the fleeces of wild and
uncultivated sheep, and wherever they abound, the value
of the wool is reduced accordingly.   In some of these
kempy fibres the usual curled or waved character of the true
wool is replaced by a twisting of the whole fibre round its
axis, so as to give the fibre the appearance of a corkscrew
with a comparatively wide thread.   Dr. Bowman has observed all the variations in the fibres from the same sheep
in the various races which inhabit Central Asia; whilst in
most of the sheep inhabiting other parts of the world, the
usual variations from the normal types are less destructive in
their character, and confined within narrower limits.   This
seems to point to the mountainous regions of Central Asia

as the district from which the present domestic sheep has spread over the other countries of the world.

Wools for commercial purposes are divided into short-stapled, or carding wools, used for woollen cloths, and long-stapled, or combing wools, used for worsted stuffs.

The British breeds of sheep yield wools which are classed indeed as long and short, or long and medium, but all, except a portion of the shorter kinds, are used for combing purposes. They are conveniently divided into two general classes :—

I. *Long-stapled wools*, which include the following :—(1.) The *Lincoln*, obtained from what Youatt thinks may be considered the parent of the English long-woolled breeds, and which is at least a good type of our deep-grown wools; it has a long, bright, silky staple, which makes it well suited for lustre goods, in imitation of mohair and alpaca textiles. (2.) The *Romney Marsh* wool, also obtained from an ancient breed; like the Lincoln it is somewhat coarse, but long, lustrous, and adapted for the same kind of goods. (3.) The *Leicester* wool, which is now much better than that obtained in former years; although finer than the Lincoln in the staple, it is not so soft and silky, yet highly esteemed for combing purposes. (4.) The *Cotswold* wool, which resembles the Leicester, but rather harsher. (5.) The *Blackfaced* breed yields what is rather a wool of middle length than long, yet it is usually classed with the long-stapled wools. In quality it varies much, according to the care bestowed upon it, but its general use is confined to rugs, carpets, and Scotch blankets.

II. *Short-stapled wools*, which are principally furnished by the different breeds of Downs. These all possess the same general characters, so that any differences which do exist are owing more to soil and climate than to any particular variety of breed. (1.) The *South Down* is a short-stapled wool, fine in the hair, but slightly harsh and brittle, of which the longer qualities are combed—*i.e.*, prepared for worsted yarns—and the shorter for making flannels and light woollen goods. (2.) The *Hampshire Down* only differs from the last in being usually longer in the staple and somewhat coarser. (3.) The *Oxford Down* is still longer and coarser. (4.) The *Shropshire* has a longer and more lustrous staple than any of the other Down breeds, and is increasing in importance. (5.) The *Cheviot* wool is of medium length, fine haired, and largely used for both worsted and woollen goods, fine cloths excepted. (6.) *Welsh* wool has some of the fine qualities of the South Down, but has a hair-like texture, and is deficient in the waved or spiral form, which gives a special value to high-class wools. Much of it is consumed in the manufacture of flannels and blankets. (7.) *Shetland* wool is not unlike the Welsh in most of its properties, only it is much finer and softer. Its exceeding fineness is well seen in the beautifully-knitted shawls, veils, and the like, made by the peasantry of the islands.

For technical purposes fleeces are divided into two classes, according to their age; thus the fleeces shorn from hoggs or tegs, as a rule, are longer and more pointed and spiral in their staples than those clipped subsequently from the same sheep. This yearling wool, accordingly, possesses properties which suit better some class of goods than that obtained from older animals, and fetches a higher price in the market. The other class of fleeces comprises those obtained from sheep in their second and subsequent years, and generally a distinction is also made between ewe and wedder wool. In some districts in the South the farmers clip the lambs, in which case the fleece is called shorn lambs' wool. There is yet another description of wool known as skin wool, which is obtained from sheep slaughtered for food. As this is obtained at all seasons of the year, it varies much in character, even from the same breed of sheep.

Packing.—In respect to the packing of wool after it is taken from the sheep's back, farmers cannot exercise too much care and prudence. It is well known that a very large proportion of home-grown clips are most carelessly assorted and stored during the short time they remain in the farmer's hands after being shorn. The first step in the management of wool is taken before the sheep are even washed. This consists in properly tagging all the dirty wool off the sheep before putting them into the wash-pool. Again, between washing and shearing, the whole flock requiring it should be carefully tagged. At clipping time, as soon as the fleece falls from the shearer, it should be immediately carried to the winders, who, before rolling it, should cut away with shears provided for the purpose every particle of soiled wool that may be adherent to the hinder parts of the fleece. When this is done, the proper way to roll up the fleece of a blackfaced sheep is to spread it out full length skin side uppermost on a table with an open top to allow sand and dust to fall through on to the ground. One of the two winders required then twists a band for tying the fleece out of the neck wool, without detaching it from the main body ; and while doing so, the other assistant folds in the edges of the fleece, and then begins to roll the wool from the tail end, pressing it moderately firm, until the neck is reached, when the operator with his band ready winds it around the whole, and binds it securely. Whitefaced wool is winded in the same way, only with this difference, the skin side of the fleece is kept on the outside. No dirt or loose wool should be rolled up in the fleece. The "gatherings" and clippings should be packed by themselves, and sold separately.

It is a common practice among sheep farmers to carry the wool direct to the wool-house or barn, and there pile it up, keeping the hogg, ewe, and wedder fleeces distinct from each other, to await the inspection of buyers who travel around the country for that purpose once a year. Those not having suitable accommodation on the farm for storing wool in a loose condition, generally pack it in sheets on the spot of clipping, and afterwards consign it to the brokers for sale. In every case wool requires to be kept in a cool condition. High temperature causes fermentation, which quickly destroys that valuable quality in wool called lustre. It is desirable on that account that wool should be treated at low temperatures. It is found that wool washed in cold water is bright, but becomes lustreless when washed in warm water. Some of the sheep dips in use also tended to make wool lose its lustre. This question of temperature has, doubtless, had much to do in framing the popular saying :—

> " Keep oo', and it will be dirt ;
> Keep lint, and it will be silk."

Sale of Wool.—In selling wool the farmer has a choice of various systems. He can either dispose of it at home to the brokers who make a practice of visiting farms immediately after the sheep have been clipped, or he may bargain with any of the buyers who attend the annual wool fairs held at different centres throughout the country. Either of these plans offer an opportunity of receiving a fair market price for the commodity in question. Sometimes, however, a buyer cannot always be found who is willing to give as much for the wool as the farmer thinks it is worth. In that case the seller must either be content to keep the wool in his own hands for a time, or else consign it to some firm of wool brokers who sell on commission. The advantages of this system are, that so long as the wool remains unsold it is free from risk, being fully insured in the agent's hands; and, further, an advance can generally be obtained from the broker, although the wool need not necessarily be sold until such time as the farmer directs. It may be mentioned, however, that the charges and interest on accommodation of this kind are not by any means trifling, as the brokers cannot consistently undertake such work and risk at nominal fees.

The sale of wools by commission is a system which has assumed, within a comparatively recent period, enormous proportions. The bulk of the wool grown in this country, and also that which is imported from abroad, nearly all passes through the commission agents' hands. Buyers, doubtless, find it more convenient for their part, and for this reason perhaps, more than any other, a generally better price can be obtained than when the wool is sold by private bargain. Farmers, however, are not altogether satisfied with the secrecy of the wool-brokers' profession, and only very recently a movement was started by some of the most extensive stock owners in Scotland for reform in connection with the sale of wools by commission. In a circular issued on the subject it is said :—

"All secrecy should be abolished, for when there is such in the conduct of business there is a tendency to a suspicion that all is not straight. The consigner should know without fail when his wool is sold, and should be able to verify the price. The buyer should know the name of the farm on which the wool has been produced—an important aid to him in deciding its value, as his examination of it in sheets can only be partial, and the salesman should have his benefit by the large extension of his business caused by an improved system. The salesman may argue that so much knowledge is more than would be beneficial to the trade, as it might tend to bring buyers and producers into direct communication ; but such arguments are not supported by experience, for like absence of secrecy has not had this effect in regard to live stock auction marts ; on the contrary, has conduced to a much greater extension of such sales than has been attained to in the sales of home wool. We, the undersigned, are agreed that the following rules would have a good effect, viz :—

" Rule A. That before each sale a catalogue of the wools for disposal should be printed and circulated by the salesman, giving such name and designation for each clip as consigners may direct.

" Rule B. When the sale is concluded, a notice to be sent to each consigner by the salesman, giving particulars of any lots of his (the consigner's) sold, and the name of the purchaser.

" In our opinion the good effects would be :—

" 1st. The consigner would with more certainty receive the proceeds from his wool.

" 2nd. The buyers would with more confidence be able to give full value.

" 3rd. The producer would have an interest in getting a good character for his wool.

" The salesman would receive for sale a larger quantity of wool, and in better condition.

" Therefore, we, the sellers of wool, inform the wool salesmen that in future we will only consign wool to such salesmen as will agree to these rules."

In relation to this circular a meeting of the sheep farmers of the North, assembled at the Inverness Sheep and Wool Fair, 1887, was held to discuss the present system of disposing of wool through brokers. The following is a newspaper report of that meeting, which it is well to preserve for future reference :—

" Mr. Mackay, Melness, Chief Magistrate of Thurso, who was called upon to preside, said they did not meet in any hostile spirit to wool-brokers, but they felt that the time had now come when the question of the sale of wool by commission must be put upon a more satisfactory footing than it at present occupied. He thought they had departed for many years from the good old system under which they had

sold their wool. In former days, when he began business
in connection with the wool trade, catalogues used to be
sent them at the time of the sales, in which useful information was given with regard to the quantities sold in the
market and the position of their own clips. That was one
of the things they now aimed at. He hoped they would
have a free and frank discussion on the question, and so
arrive at a satisfactory conclusion.

"Mr. Scott, Alnwick, said for some time there had been
a good deal of agitation in the minds of farmers and wool
producers with regard to the way in which the sales of wool
by commission had been conducted. They were dissatisfied
with the principle upon which the wool-broking business
was at present carried on; and at the request of a large
number of the biggest sheep farmers in the country, he had
carried on an extensive correspondence, the result of which
had been, that he had received favourable replies from
almost every one, with the exception of a few wool-brokers,
who did not approve of the proposals made for remedying
the present state of matters. He trusted, however, that
farmers would unite together, and that, as a result, they
would be successful in obtaining the reform they desired,
and that the wool-brokers would also agree to their proposals.

" Mr. Thomas Purvis, Rhifail, Caithness, moved the first
resolution as follows :—' That in the opinion of the meeting
the conditions under which wools are at present sold by
commission are most unsatisfactory, both to the buyer and
the seller, and reform is necessary.' The motion was unani-
mously adopted.

"Mr. Gordon, Balmuchy, proposed 'that the meeting,
having taken into consideration the terms of the circular
which has already been published, adheres to and endorses
the same.' The motion was, after some discussion, unani-
mously agreed to.

" A committee was appointed to consider any proposals

P

wool-brokers may make, and to intimate to the signatories if such are considered satisfactory—Mr. Scott, Alnwick, being convener."

Commenting on the proceedings at the above-mentioned meeting, Mr. Ramsden, of Russell & Ramsden, Leith, said— "Some of the rules suggested for the conducting of wool-broking were in use twenty years ago, but were discarded by the wool growers as not being in their interest, while some of them were quite new, and from a wool-broker's point of view appeared to involve commercial *felo de se*, and therefore they ought to object to them. The interests of wool-brokers and wool-growers were intimately connected. It was the interest of wool-brokers to endeavour to realise the most that could be made for the wool consigned to them, and if the influential body of wool-growers who were promoting these reforms would propose something that really, and not apparently, was in their own interest, and did not involve commercial extinction to the broker, the firm which he represented would be the first to adopt such changes."

The opinions expressed at conferences have not yet had time to operate, but it is believed that conciliatory arrangements between buyer and seller will by-and-by be adopted.

Wool and Speculation.—The question has not infrequently presented itself—Should farmers be speculators in farm produce, and, if so, to what extent? It is observed in the wool trade that the farmer is most willing to force his wool on the market when prices are low, fearing further flatness. With high prices he holds for a rise. There can be no possible objection to any farmer who has more capital than he finds necessary for the purpose of carrying on his holdings speculating in his own grain or wool, and even extending his operations as far as prudence may dictate. In this case, however, stock should be taken, and the farm accounts regularly squared and kept distinct from his speculative transactions. This may seem an unnecessary and uncalled-for observation, yet occasionally it should be salutary advice. The proceedings in our bankruptcy courts at times show that the blame of a farmer's failure does not always rest with his farm, and that his severe losses were caused by a heavy depreciation in the value of produce which he had stored up in expectation of a rise in the markets, which anticipations, unfortunately for him, were not realised, and the consequence was a disastrous result.

A year or two ago it was far from an uncommon occurrence to be told that one farmer had six or seven clips, and another as many as twelve or thirteen clips stored away somewhere about his farm steading. These men were holding on for a rise which has never come, and the consequence was, that clips which in their year would have brought 40s. had in the end to be sold for half that sum. When wool was 40s. or even 30s. they were too grasping to accept the then current price ; wool went down to zero, and when they were forced to sell they had to pay sweetly for their whistle. They not only lost an enormous amount of capital, but also the interest—in cases where the former might have been invested, compound interest. Their original capital then has dwindled down very considerably. They are crippled in their resources, and all this is too frequently "laid at the door" of their farms.

As a rule, the farmers who have sold their wool every year have been all along in a much better position than the speculators ; and at the present moment, even with wool at its minimum value, it would be the reverse of wisdom for farmers to hold on their clips in expectation of a rise. Wool will, undoubtedly, rise in price some day, but that period is not by any means within the range of near probability, and until it does appear speculation had best be avoided.

CHAPTER XXV.

HIRSELS AND HERDING.

THERE is no word more common in the shepherd's vocabulary than *hirsel.* It has several meanings, however, and is written in various ways. Jamieson gives it as *hirsell, hirdsel, hirsle,* and *hissel,* meaning a multitude, a throng, applied to living creatures of any kind. Ramsay spells it in two different ways—" Ae scabbed sheep will smit the hale *hird-sell.*" Again—

> " Near saxty shining summers he had seen,
> Tenting his *hirsle* on the moorland green."

Burns sings—

> " The herds and *hissels* were alarmed."

Another derives it from Fr. *haraz* or *harelle ;* sax-*herd,* grex. The term is by no means restricted to a flock. A drove of cattle is, indeed, called a *hirsel* of beasts ; but it is common to speak of a *hirsel* of folk, a *hirsell* of bairns. If we suppose that it was primarily applied to cattle, Jamieson thinks that the first syllable may be derived from Su.-G. *haer,* an army, and *saell-a,* to assemble ; whence *saell,* a company, a company assembled, which precisely expresses the general idea conveyed by the term.

It is also used by Hogg in *Brownies of Bodsbeck* as expressive of quantity or number. " Jock, man," he says, " ye're just telling a *hirsel* o' e'endown lees " (lies). But Scott uses it with an entirely different meaning :—" So he sat himsel' down and hirselled down into the glen, where it wad ha'e been ill following him wi' the beast" (*Guy Mannering*). "The gude gentleman was ganging to *hirsell* himsel' down Erick's steps, whilk would have been the ending of him, that is in no way a crag's-man" (*The Pirate*). In this sense hirsel means to rub, or move forward under difficulties. As applied to sheep, the word is exceedingly appropriate and expressive of its meaning. The flocks on large farms are divided into hirsels or companies sufficient for one shepherd to tend. Thus, in asking the size of any particular farm, people say, "How many hirsels are on it?" referring to the number of shepherds required, and not to the extent or acreage of the holding. *Hirst* is another word of similar origin, and means a barren height or eminence ; but in some districts it is used to signify a large number, as "a hirst of weans."

Hirsels are in a manner separate and distinct farms in themselves. Each hirsel has a shepherd, and although undivided by fences, the flocks of each are quite distinct and keep to their own ground. Trespassing is as rare between the sheep of different hirsels as it is between those of different farms. The natural conformation of the hills has assisted in the division of the farms into hirsels, the boundaries of which are usually marked by mountain-streams, or height and hollow. The arrangement of the hirsels, long ago accomplished, is seldom if ever altered, which speaks volumes for the sagacity of our forefathers in sheep management. The slope and character of the ground have been so well surveyed that, if there is any one thing permanent about a sheep-farm, it is the "run" of the sheep, or way in which they have been taught to travel over the ground.

Very much depends upon how the sheep are guided over the pastures. Were they left to their own free will the herbage would be unequally fed, and the consequence

would be that the sheep would literally starve one another
by grazing continually on the sweeter parts and leaving the
coarser untouched.   To prevent all this the sheep on each
hirsel are divided into "cuts" or bands, and each cut has
its own range or feeding ground, which is so apportioned
that the sheep it carries receive a share of as much variety
of soil and grasses as possible.   The number of sheep in
each cut varies according to the description and slope of
the ground.   On some farms, where the surface is even and
regular in outline, the sheep in each cut may number as
many as ten or twelve scores, whereas when the ground is
irregular and difficult of access, from two to five scores are
about the average size of the cuts.

An important point in herding is to divide the sheep as
equally as possible over the whole of the ground.   It is as
easy to overstock a heft as it is to overstock a farm.   As a
matter of fact some shepherds are sadly unconscious of the
great error they make in arranging the number of sheep
suitable for each cut.   Should one cut contain more than
another according to privilege of pasture, there is certain to
be a difference in the quality of the stock raised on each.
And as soon as one cut gains an advantage from being lighter
stocked, it is not difficult to foresee what will happen at
weaning time when the best of the ewe lambs are selected
for keeping purposes.   The lambs which are retained to
keep up the stock invariably find their way back to the heft
on which they were bred; and it is evident that the best-
conditioned cuts will contribute more than their share of
lambs kept for that particular year.   It therefore follows
that the lightest stocked cut this year is apt to be heaviest
the next, and while the distribution in a manner equalises
itself, it is a process too vague to be productive of the
greatest possible benefit to the stock.   Shepherds who are
good at "kenning" their sheep by headmark can select the
stock lambs with great accuracy, so as to avoid overstocking

any of the hefts; but the only sure method is to collect and draw each cut of sheep separately. This involves a good deal of trouble, and is perhaps never carried out; still that is no reason why it shouldn't. At all events, any one who really wishes to farm upon a scientific principle will find it to his material advantage to carry out this system in practice.

Cuts.—The size of the "cuts," or number of sheep which should be allowed to range together, is a question on which the health of the flock largely depends. Change and variety of soil and grasses is the life-preserver of all grazing stock; and with hill sheep the difficulty is to obtain these in sufficient quantity. Where the pasture is very much similar in character, it seldom contains all the elements necessary for the proper development and nutrition of the animal system, and as a consequence the sheep become impaired in health and body. To obviate this evil, some flockmasters adopt the system of making all the sheep that are under one man's charge travel over the whole of their ground daily. This gives every sheep as much variety of pasture as it can possibly obtain, and the results are invariably satisfactory, especially on farms naturally unhealthy. Still, there is a limit to this practice as well as to most others. It is an axiom in sheep management to herd as few together as possible, and the system of running the whole flock in one band is obviously contrary to this rule. That is one objection to large cuts; and another is that the sheep destroy more pasture by tracking and trampling the ground more severely than if kept in smaller companies. Too much exercise is also detrimental to the accumulation of fat on an animal, and it is quite possible to injure sheep by making them travel longer distances than they can daily accomplish with ease. Small cuts are far more general than large ones; but should there be any disease, such as pining or vanquish, on the farm, then a wide range of pas-

ture is certainly to be commended.   Blackface sheep can
be herded together in larger bands than almost any other
breed, and by nature they are so constituted that several
miles of daily travel agrees with them better than confine-
ment to narrow bounds.

Herding.—In herding hill sheep, the universal practice
is to make all the various cuts meet every evening at the
highest part of the ground, where they rest for the night.
From this point each company starts off in the morning to
their several hefts.   They make no mistakes, and will separate as correctly as if they had never been mixed.   After
rising, they do not commence to feed slowly down the
slope of the hill, but string away one after another, never
halting until they arrive at the lowest part of the grazing.
Here they commence to feed until about noon, when they
again begin to move outwards.   They scatter themselves
over every part of the ground, and, gradually moving upwards, they reach the heights about dusk.   In the long
days of summer they lie down to rest once in the forenoon
and once in the afternoon; and as the days get shorter
they only rest once, about mid-day.   The shepherd is careful to see that every sheep performs this daily march over
the ground, not allowing the lazy ones to stop half-way or
monopolise the sweeter spots of the grazing.   The secret
of good herding lies in guiding the sheep evenly over
the ground.   Merely driving the sheep down and up the
hill means nothing.   Any brainless mortal could do that
much.   It is in guiding the sheep so as to give them the
full benefit which the pasture affords at all seasons of the
year—saving certain portions which will be more valuable
at another period, and consuming that which is only of
value for the present—that good results are obtained.   In
this matter there is an immense difference in what different
hepherds can accomplish for the good of the flock.   A man

who thoroughly understands how to depasture his ground both summer and winter will bring out his sheep in very superior condition to one who takes no thought in the matter, or does not rightly understand the nature of his grazing. This is one of the reasons why the changing of shepherds is to be deprecated, as a stranger takes some years to acquire the requisite knowledge respecting the geological influences of a new district.

A very remarkable point in the practice of hill herding is that of driving the sheep to the hill-top at night. Any one not acquainted with the business may be apt to think that the top of a hill would be the coldest and most exposed situation the sheep could possibly find, and ask why not give them better shelter at night by keeping them on the bottom? In a certain sense shelter is very desirable, and there is no doubt that the top of a hill is not where it is most likely to be found, but the advantages derived from a high and dry lair are more than equivalent to the evils of the same. Sheep naturally prefer to rest on the highest ground, although they can be taught to lodge on the lowest, to which they will draw in the evening of their own accord. Yet where sheep are kept in ill-ventilated, close surroundings, they are never so healthy as when they are allowed plenty of pure air, even if it is both colder and more abundant. If sheds were to be erected on the hill tops and the sheep housed every night it would not be good for them. They experience occasional blasts from which they might escape if laired on the low ground, but taking the year throughout the hill top is the best and most suitable place for them to rest overnight.

The Shepherd.—It almost goes without saying that the life of a hill shepherd presents little or no variety, either in regard to its occupations or its social aspects. He not only seldom or never mingles with the busy populations of the large cities and towns, but is a comparative stranger even to the daily routine of ordinary village life. Removed from the busy haunts of men and living in the solitudes of the mountains, he pursues the even tenor of his way with the same placid unconcern as if the big and busy world did not exist. If it is true that that nation is happy which has few annals, it is emphatically the case with the hill shepherd, whose life, as a whole, is spent in unbroken uniformity and in the enjoyment of uninterrupted happiness. But as all rules have their exceptions, so there are certain events and circumstances in the life of the hill shepherd which break, to some extent, the monotony and impart a little freshness and stimulus to its otherwise commonplace character. There is the clipping season, which takes place in midsummer, and afterwards the time of haymaking—both of which, but especially the latter, call for more than ordinary exertion, and take him out of the daily routine of his life. Then during the winter, should severe snowstorms occur, all the energies and ingenuity of the shepherd are called into requisition and taxed to the utmost in rescuing his fleecy charge from the pitiless and often fatal snowdrifts.

Shepherds occupy an important place in the general economy of Scotland. Their occupation may be almost said to be hereditary. Living in almost total solitude among the hills, they have few opportunities or inducements to spend money, and their wives are generally as frugal as themselves. A shepherd has two great objects in life—to save and to give his children a good education, for which he is willing to make any sacrifice. The abodes of these mountain shepherds are often in the most desolate and dreary situations that can be conceived. For weeks together they do not see their employer's face or hear anything of the outer world. Generally well educated themselves, they devote their winter evenings to knitting and reading.

Physically they are tall and wiry, but seldom corpulent.

They dress in a home-made suit, Scottish tartan plaid, and a few of the old hands stick to the once universal Kilmarnock bonnet. When not otherwise employed with their hands in the house or going their rounds, knitting is their favourite occupation. Some of them have acquired, no doubt from travelling on the rough hills, a peculiar swinging movement when walking; these take huge steps, and their bodies roll dreadfully from one side to the other. Suppose one marching in his natural state in a line of infantry his head would alternately nearly knock the shoulders of his right and left-hand comrades. This peculiarity is more perceptible in the younger ones. It may be worthy of remark that they mostly use tobacco, which they chew as well as smoke, and those who do not use it thus carry the snuff-box; and, though by no means a drinking class, few of them are Good Templars. While scarcely a single instance could be given of a shepherd who does not use one or more of these luxuries, there are at the same time very few indeed who abuse them. Perhaps once or twice a year they go a little too far when they come down from the hills to attend local fairs.

In a general way they are placed at great inconvenience with regard to the doctor, the church, the post-office, and the school. Some of them who live at long distances from the doctor have acquired a homely knowledge of medicine, which they apply with tolerable success on an emergency. They attend church regularly, many of them walking ten miles to do so, and in weather when some people within half a mile from the church deem it too stormy to venture out. Of late years the postal accommodation has been very much extended in Highland districts; but there are still many that do not enjoy this convenience, and letters for these will often lie at the post-office in the neighbouring village, uncalled for, from the Monday till the following Sabbath, which is made a convenient day for passing them on these out-of-the-way places. The education of their children is perhaps the greatest difficulty and expense they have had to contend with. The scanty population in many of these outlying districts did not warrant the establishment of schools, and so the herds had either to board their children with strangers in the vicinity of a school or combine among themselves and employ a teacher to itinerate among them, thus incurring an outlay they could not easily afford ; and even since the Education Act of 1872 has come into operation, School Boards experience considerable difficulty in dealing with the educational wants of those districts.

The following testimony of the character of hill-shepherds in Roxburgh and Selkirkshires is from the fourth Report of the Royal Commissioners, issued in 1870 :—

" The hill-shepherds are usually a superior class, both as to education and morals. They are particularly well up in theological subjects."—*Dr. Anderson, Provost of Selkirk.*

" The hill-shepherds are the most respectable and best-behaved class of men to be met with. Their employers place great confidence in them, and generally retain the sons to succeed to the fathers' employment. They read to a considerable extent, and are very intelligent."—*Rev. J. Falconer, Minister of Ettrick.*

" I would characterise the hill-shepherds as being among the most shrewd and intelligent men in the country. I can testify to the sobriety and the responsibility laid upon many of them, and the honourable way in which, for the most part, they fulfil their trust is highly praiseworthy. Some of them write business letters which, for style and penmanship, could not be surpassed. There are, of course, exceptions ; but as a whole I should say the shepherds in the south of Scotland are a very superior class of men."—*C. K. Green-hill, Minister of Roberton.*

Shepherds' Wages.—In the south-east of Scotland the

hill-shepherds are mostly paid in stock; that is to say, they have a "pack" of sheep numbering from forty to fifty, according to the quality of the grazing. Besides his pack, which grazes promiscuously among the farmer's sheep, the shepherds are allowed a cow's keep, sixty-five stones of oatmeal, one thousand yards of potatoes, and a free cottage and garden. The value of the wage depends greatly upon the price of wool and mutton, but generally it is computed at from £45 to £50. The shepherds are proprietors of their packs, so that, together with the value of their cow, pig, and perhaps a few chickens, they are men of considerable means.

In Ayrshire the shepherds are not paid by packs, but receive generally from £16 to £18 a year, and in addition they have pasture and hay for two cows, ten bolls of meal, a cartload of potatoes, fuel carted, and a cottage and garden rent-free. The profit from two cows is estimated at from £16 to £18 a year. This class of men maintain the same high character in Ayrshire as in other parts of Scotland, being noted for their intelligence, honesty, and frugality, and for the sacrifices they make for the education of their children.

In Lanarkshire the pack system of payment is now very little in vogue, owing to the natural suspicion, whether well-founded or not, that the shepherd would take more care of his own sheep than his master's. Where packs are allowed in this county they number from forty to forty-five sheep. The other payments consist of half an acre of potato ground, sixty to eighty stones of meal, keep of cow, and the shepherd cuts his own peat, in which work he is assisted by his own family or a hired boy. A money wage in Lanarkshire is about £30 per annum, given as an equivalent to the sheep, the other gains being the same as in the former. Where shepherds have a boy to keep as assistant, who boards with them, the keep of another cow is allowed, four cwt. of potatoes, and sixty-five stones of meal. The boy gets about £12 in wages, which is paid by the farmer.

In the West Highlands the shepherds are allowed part stock and part money as wages. A Cantyre shepherd puts his wage down at £20 in money, grass for two cows and a calf, nine bolls of meal, potato ground planted, free house and garden, sheep who die of braxy, and whisky at clippings. Most shepherds have also one or two fat sheep. The pack system was general at one time, but is now seldom adopted as a whole. The keep of a cow is valued at £3 in Argyllshire, but it varies, of course, according to the richness of the pasture; in some places where the land is very rough it is worth very little. An extensive farmer writes that he pays his shepherd £12 in money, two cows' grass, two stirks' grass, seven bolls meal, potato land, house and garden, and braxy, or fallen sheep.

In the Highlands both pack and money-wages are paid, but the latter most generally. Shepherds with packs have not had nearly so good a wage during recent years, owing to the low prices for sheep and wool, as those with a money equivalent. The money-wage is certainly the safest for the shepherd, although when sheep were dear the farms which allowed a pack were paying a great deal the highest wages. On the other hand, shepherds with packs have been having very poor wages for a number of years recently, and many of them are heartily tired of the system, and striking for money payment or an augmentation to the number of sheep. At the present prices for sheep it takes a very large drove of lambs to pay the shepherd's wages.

The Shepherd's Crook.—The crook is a very useful implement for the shepherd, and with it a moderately gentle sheep can be easily caught from a number huddled into a corner, without a man springing among them and frightening all the rest. It should be used with a quick and gentle motion, and the caught sheep immediately drawn back rapidly enough to prevent it from springing from one side to the other, thus wrenching its leg or throwing itself down, by exerting its force at an angle with the line of draught. The crook should be placed immediately above the hock in catching the sheep. When the sheep is drawn within reach, the leg held by the crook should at once be seized by the hand, and the crook removed.

The crook had several uses among ancient shepherds. The straight end answered for a cane. It was also used as a weapon of defence against robbers and wild beasts while guarding their flocks. But perhaps there was no more common use than that of guiding and controlling the sheep. With the hooked end the shepherd caught and punished the turbulent ones. Among European and Oriental shepherds these crooks are still met with, though with the Europeans they are often made of solid or hollow iron. Bishops in the Christian Church, at quite an early day, took the name of shepherds, and carried a symbol of authority called a crosier or pastoral staff, which was quite similar to the ancient shepherd's crook; and when one of them neglected his duties it was said of him—

" He left his *crook*, he left his flock."

Strange to say, the old-time symbol of authority is now seldom seen, unless among the rubbish or scrap-iron heap at farm sales. Modern shepherds, who are sometimes spoken of as a wee lazy, have devised a different kind of crook, which serves the double purpose of a walking-stick and crook combined. The staff of an ordinary walking-stick being too short for catching a sheep by the leg, the crook has been widened so as to enable the shepherd to catch his sheep around the neck. For many purposes the neck-crook is much to be preferred; yet the old leg-crook was an implement that saved an immense deal of labour and abuse, especially to sheep stanced about open fairs and markets, where a good deal of mixing and catching-out was the rule. There should at least be one long-handled crook always handy at sheep-sortings, as its use will save many an unnecessary plunge by the gripper, who not only runs the risk of straining himself, but often inflicts serious injury on many of the sheep.

CHAPTER XXVI.

THE COLLIE.

"The colly dog lies i' the nook,
  The place whilk his auld father took,
  And aft toward the door does look
    Wi' aspect crouse,
  For unco folk he canna brook
    Within the house."

THE COLLIE STANDARD.

THE standard, as drawn by the English Collie Club at the Kennel Club's annual show in July 1885, is as follows:—

The skull of the collie should be quite flat and rather broad, with fine tapering muzzle of fair length, and the mouth the least bit overshot, the eyes widely apart, almond-shaped, and obliquely set in the head; the skin of the head tightly drawn, with no folds at the corner of the mouth; the ears as small as possible, semi-erect when surprised or listening, at other times thrown back and buried in the "ruff."

The neck should be long, arched, and muscular; the shoulders also long, sloping, and fine at the withers; the chest to be deep and narrow in front, but of fair breadth behind the shoulders.

The back to be short and level with the loin, rather long, somewhat arched and powerful. Brush long, "wi' upward swirl" at the end, and normally carried low.

The fore-legs should be perfectly straight, with a fair amount of flat bone, the pasterns rather long, springy, and slightly lighter of bone than the rest of the leg; the foot with toes well arched and compact, soles very thick.

Q

The hind-quarters, drooping slightly, should be very long from the hip-bones to the hocks, which should be neither turned inward nor outward, with stifles well bent. The hip-bones should be wide and rather ragged.

The coat, except on legs and head, should be as abundant as possible; the outer coat straight, hard, and rather stiff; the under coat furry, and so dense that it would be difficult

PRIZE COLLIE "ROB ROY."

to find the skin. The "ruff" and "frill" especially should be very full. There should be but little "feather" on the fore-legs, and none below the hocks on the hind-legs.

Colour immaterial.

Symmetry.—The dog should be a fair length on the leg, and his movements wiry and graceful. He should not be too small. Height of dogs, from 22 to 24 inches; of bitches, from 20 to 22 inches.

The greyhound type is very objectionable, as there is no brain-room to the skull, and with this there is to be found a fatuous expression and a long, powerful jaw.

The setter type is also to be avoided, with its pendulous ear, full, soft eye, heavily feathered legs, and straight, short flag.

The smooth collie only differs from the rough in its coat, which should be hard, dense, and quite smooth.

SCALE OF POINTS.

Head and expression	15
Ears	10
Neck and shoulders	10
Legs and feet	15
Hind-quarters	10
Back and loins	10
Brush	5
Coat and frill	20
Size	5
Total	100

Note.—Point-judging is not advocated, but figures are only made use of to show the comparative value attached to the different properties. No marks are given for "general symmetry," which is, of course, in judging, a point of the utmost importance.

In commenting on the above, Mr. H. Panmure-Gordon writes:—"In my opinion, the scale of points should rank thus:—

Head and expression	20
Coat and frill	20
Legs and feet	15
Back and loins	15
Ears	10
Neck and shoulders	10
Brush	5
Size	5
Total	100 "

In attempting to give instructions for training a collie it is impracticable to lay down fixed rules, as each dog has his own characteristics. The trainer, to be successful, must study the temperament of his dog, and be guided in great measure by it. Avoid by all means having children play with your puppy. Feed him yourself, and get him attached to you. Let him go with no one but yourself, and when you do not wish him to go with you, put him in a kennel or chain him up. If the dog will not do what you wish, call him always to you; but never throw anything at him, and chastise him very gently, but never in such a way as to make him afraid of you. Many a good dog has been spoiled when a puppy by giving him an injudicious thrashing. Be firm with him, never making too free with him. Make him understand that whatever you wish him to do he must do it without question.

When a pup is between eight and nine months old it is generally the best time to begin to train him, but it depends a good deal upon his growth, and whether he has by this time developed speed enough to run past either sheep or cattle. This is very important to know, for there is nothing so discouraging to a young dog as not to be speedy enough. When you are satisfied that you have your dog so that he will come to your foot when you call him, and stand still when you wish him to do so, begin the training. You can teach him to stand still by seating him down facing you and stepping back yourself, keeping your eye upon him all the time, and saying, "Stand, Clyde," or whatever his name may be, until you get back a considerable distance, always increasing the distance as you get him better instructed, until you can finally go away and he will sit a little time till you get back; at least have him so that so long as your eye is on him he will not get up; then vary this a little by letting him run some little distance off, and then hold up your arm and cry out, "Sit down," or "Stand." By this means you will get him so that by simply raising your arm and dropping it he will lie down. Encourage him when he does right, and always reprove him when he does wrong; do not give long lessons at first, and be sure to always use the same words for the same things.

One of the first lessons to teach the dog is how to pen sheep. Get four or five sheep that are not too wild, and drive them into a small enclosure—a horse-paddock or cattle-yard will do. Drive the sheep into one of the corners, keeping the dog with you, and make him watch one side while you watch the other. Move up on the sheep until you get one to run past you, and then make the dog run and bring the sheep back to where it started from. Use the dog at this until he will do it perfectly by a mere motion of your hand; and on no account let him bite a sheep without punishing him, as a dog with this habit will soon be good for nothing.

For the next lesson teach the dog how to head sheep, or turn and bring them to you on the open space. Take your dog out into an open field, not too large, with ten or twelve sheep that have been somewhat accustomed to seeing a dog, if you have such; call him up and keep him behind you; then send him to head off the sheep and bring them towards you. The wider the circle you can get the dog to take before he heads them the better. If he attempts to run straight toward the sheep call him back and send him off again, waving and encouraging him with your hand to go wider. Cry out to him, "Back wide," "Far away," "Back wide," when he has headed them. Encourage him to bring them to you by crying, "Come on with them," and when you get them where you want them, then call the dog up to your heels—keeping him always behind you. The wider you can get a dog to run the better, till by usage you can send i m as far as you can see him by a wave of the hand. Use him to run either way by a wave of your right or left hand, as you may wish him to go to the right or left, and when he is running cry out at him, "Go wider, far away wide," if he is running too straight. Accustom him to lie down when he heads off the sheep by crying at him, "Lie down," and holding up your arm. If he should happen to run away from you altogether, go straight after him and bring him to where he ran from, and do not punish him till you get him there; then is the time to punish him, and make him do what you desire. This is one of the principal lessons that a dog needs to be taught. Never let your dog run among the sheep—always send him around them. There are many other things you may teach a dog, such as going before the sheep on a road, you driving them from behind. Going before yourself, you may teach the dog to drive them from behind. All these are easily learned by continuous practice. In fact, you would think your dog at this stage knows better what is to be done than you do yourself. Unless the dog is persistently disobedient, no severe punishment should be resorted to. Ordinarily taking the dog and placing him where you wish him, or making him assume and keep the position until released by you, will be sufficient. If any punishment is needed, a little pinching of the ears will be enough. Kicking the dog or scolding him habitually will soon spoil him. If the dog is too much inclined to bite the sheep, a muzzle may be used in the early stages of training; but this is seldom necessary. Above all things, let no one have anything to do with the dog except the trainer.

"A Shepherd" thus discourses in the *Farming World* on breaking a dog to sheep :—

"Take the dog among a flock of strong fleet sheep, and encourage him to take his will of them for a few turns. Do not say a word to him so long as he keeps their skin whole. If the dog is keen to run, less of this will do; but he ought to be allowed so much of his own way at first as to let him know you have no objections to him running.

"Once you get him to go to the front, endeavour as quietly as possible to keep him there, running where necessary yourself, to keep the sheep between you and the dog. Get him, if possible, to fix his attention on the sheep; move them forward, if possible, at the same time making him move on too. This will take with some dogs considerable time, patience, and care. He must not be rashly checked, even though he does wrong.

"With a keen, natural-working dog, the case will be a more easy one. Give him a few turns at running in front, keeping him at a respectable distance from the sheep by threatening to throw a stone or stick, or even doing so if he will stand it. You may now command him gently if he is at all interested in going ahead. I may say that I never had a dog but was keen on running ahead, more so than anything else.

"Set him to run off the hand to the front with a circle, and stand there, awaiting further orders. In no case let him bring them towards you for a considerable time; rather go to him. This is a point where many a dog-breaker fails. The dog naturally inclines to bring them in, but when allowed to do so at first, without stopping, he gets so 'good' at it that no sooner is he beyond them than he drums them in faster than they can run, often overturning one or two by the way, or causing them to creep down for want of breath.

"He will now have to learn to 'come in ahint;' that is, from your right side round behind you, and out off the left with a sweeping circle, or from left to right the same way. Care must be taken not to check the dog until he gets right round in front, unless to widen him should he come too close. It would be better to avoid this by having the sheep so near, before putting him away, that he will take them without bungling at the one sweep. He will go wider in time. In no case allow him to come between you and the sheep. Some dogs are rather annoying at this point by persisting in running only off one hand. A little patience and care overcome this difficulty.

" Those who require 'hunt awa'' dogs ought not to make a 'holding-in' dog hunt off till he is at least two years old. A dog is quite soon enough at work when ten months old. When put out too young to be able to turn a fleet sheep, he learns to chirp and bark.

" Too much of the hazel spoils many a good dog—a practice that cannot be too much condemned. A dog should know you can throw a stone or a stick, but should have no fear to come up to your hand. Young dogs are better tied up when not in use; running after carts, crows, &c., spoils them greatly.

" Finally, I would say, after a few turns to begin the dog, work no more with him but your necessary everyday work. By so doing, and keeping the dog under proper command, you will have him as well broke as though you had wrought a lot of needless work, and given the stock much annoyance, as many a young hand does."

The collie, as is well known, is a constant attendant of the Scotch shepherd, and with the help of his dog he has little trouble in handling a flock of blackfaced sheep. An educated collie can do more work than ten men among a flock of sheep. His patience and fidelity to his trust are without doubt unknown in other breeds. They can be taught to perform the most remarkable feats, but do not like to be imposed upon, and from their constant association with man, seem to attain a degree of intelligence that they are usually spoken of as being able to do " everything but speak." As pets they are lovable, and worthy of the growing popularity they are achieving. The spayed females are very tractable, and make excellent dogs for cattle or sheep, as they are generally very gentle in disposition. From the fact of the gentleness of the females, they are, when non-breeders, very desirable as pets.

There are many types of the Scotch collie, the most fashionable at the present time being the rough-coated sable; if any white, it should be in the breast and tip of tail. Black and white are also favourite colours. These dogs should be fed, especially when young, upon oatmeal cakes or brown bread and skim milk. They may have bones with a little meat on them to gnaw, but too much meat is injurious to their health and activity. The dogs when bred from ancestors that have been trained for several generations seldom fail to give satisfaction. There are many collies purely bred but almost worthless. Like every other animal, to be good they should be descended from stock that has proved its worth.

For almost any purpose required of dogs the collie has no peer. He will protect his master or his property even unto death, and his faithfulness of trust far exceeds that of average humanity.

A good story of the sagacity of the collie is told by Mr. Louis R. Stephenson of one John Todd, a shepherd on the Pentlands, near Edinburgh. This worthy had been a shepherd all his days, but he declared that he only possessed one dog in his sixty years' experience that was really gifted with superior wisdom. He had been offered £40 for it. Once, in the days of his good dog, he had bought some sheep in Edinburgh, and on the way out, the road being crowded, two were lost. This was a reproach to John and a slur upon the dog; and both were alive to their misfortune. Word came after some days that a farmer about Braid had found a pair of sheep; and thither went John and the dog to ask for restitution. But the farmer was a hard man, and stood upon his rights. "How were they marked?" he asked; and since John had bought right and left from many sellers, he had no notion of the marks. "Very well," said the farmer, "then it's only right that I should keep them." "Well," said John, "it's a fact

that I canna tell the sheep; but if the dog can, will ye let
me have them?" The farmer was honest as well as hard,
and, besides, he had little fear of the ordeal; so he had all
the sheep upon his farm into one large park, and turned
John's dog into their midst. That hairy man of business
knew his errand well; he knew that John and he had bought
two sheep, and (to their shame) lost them about Borough-
muirhead; he knew, besides (the Lord knows how, unless
by listening), that they were come to Braid for their recovery,
and without pause or blunder singled out first one and then
another—the two waifs. It was that afternoon the £40
were offered and refused, and the shepherd and his dog, or
rather the true shepherd and his man, set off together by
Fairmilehead in jocund humour, and "smiled to ither" all
the way home, with the two recovered ones before them.
So far so good; but intelligence may be abused. The dog,
as he is by little man's inferior in mind, is only by little
his superior in virtue; and John had another collie-tale of
quite a different complexion. At the foot of the moss
behind Kirk Yetton (Caer Yetton wise men say) there is a
scrog of low wood and a pool with a dam for washing sheep.
John was one day lying under a bush in the scrog, when he
was aware of a collie on the far hillside, skulking down
through the deepest of the heather with obtrusive stealth.
He knew the dog—knew him for a clever rising practitioner
from quite a distant farm—one whom, perhaps, he had
coveted as he saw him masterfully steering flocks to market.
But what did the great practitioner so far from home? and
why this guilty and secret manœuvring towards the pool?—for
it was towards the pool he was heading. John lay the closer
under his bush, and presently saw the dog come forth upon
the margin, look all about to see if he were anywhere observed, plunge in and repeatedly wash himself over head
and ears, and then (but now openly and with tail in air)
strike homeward over the hills. That same night word was

sent his master, and the rising practitioner, shaken up from where he lay all innocence before the fire, was had out to a dykeside and promptly shot; for, alas! he was that foulest of criminals under trust, a sheep-eater; and it was from the maculation of sheep's blood that he had come so far to cleanse himself in the pool behind Kirk Yetton.

CHAPTER XXVII.

ANNUAL MORTALITY AMONGST SHEEP.

SOME one has said that "sheep will, and must die," and in one sense the saying is true; but it is not true that sheep *must* die prematurely of disease, mismanagement, hunger, and exposure to severe winter weather in the immense numbers they now do. The sheep is naturally a hardy animal, singularly free as animals go from constitutional ailments, and subject to very few contagious diseases, yet the average annual loss amongst hill flocks in ordinary seasons cannot, perhaps, be put at less than 10 per cent. Even on some lowland farms, where sheep management was not well understood, we have known this percentage to be exceeded. And when long and severe winters occur, as they very frequently do, it is by no means exceptional for hill flocks to suffer to the extent of 15 or 20 per cent.

It would lead us too far afield to go into details of sheep losses from the various diseases, and other causes, but it is generally admitted that they are enormous. The extent to which the sheep stock of this country have suffered from foot-and-mouth disease alone may be judged from its frequent recurrence in past years, and the fact that during many of the outbreaks as many as 25 and even 50 per cent. of the sheep have been attacked. Again, there have been four serious epidemics of liver-fluke within the last forty years—in 1853, 1860, 1869, and 1879, those of 1860 and 1879 being the most disastrous. In 1860 the loss of sheep from this cause is believed to have been close upon 2,000,000, and in 1869 it must also have been considerable, for the returns of the following year show a decrease of 1,500,000, one-third of the deficiency being in lambs. Mr. J. L. Bowes, of Liverpool, puts the United Kingdom loss by liver-fluke in 1879 at 2,889,000 sheep, besides a deficiency of about 2,400,000 lambs the following season, making a total decrease in our flocks of 5,250,000 sheep from this one cause. An inconceivable amount of nonsense was spoken and written on this subject of "sheep-rot" during the continuance of the last epidemic. Quackery is bad enough in such matters, but ignorance on points of simple fact bearing on the same is even more inexcusable. We remember two long letters in the *Times*, descanting on the epidemic, gravely told its readers of the wisdom of the people of Westmoreland in keeping flocks of geese to eat up the snails ; as if any one ever saw a goose eat a snail !

That many of the losses which occur amongst sheep are preventible, none conversant with the facts will deny; but no material improvement can be expected until the necessary data for correction are supplied. What requires first to be made known in this connection is the different causes of sheep losses, and the percentage of loss attributable to each cause ; also, how far the general loss, or particular causes of loss, are influenced by feeding and management, geological formation of soil, wet or dry seasons, &c. The only way to arrive at this is for farmers to make a register of all their sheep losses.

A Scottish farmer, who has kept a register of this kind for some years, writing to us on the subject in the beginning of last year, says :—" I make my head shepherd insert in his monthly report of the flock the cause of all deaths, losses, and sales from delicacy, &c.; and I have tabulated the results for twelve months ending October 31st, 1885, showing that out of a sheep stock of about 2000 we have during the year lost by death and compulsory sale 167, or rather more than 8 per cent. I am very far from thinking that the register is otherwise than crude and imperfect; but I venture to send it, in the hope that it may be of use to you in advocating what I agree with yourself in thinking is an important subject to the stockmasters of Great Britain. On October 31st the actual total number of sheep in the flock was 2033, consisting of 1210 breeding ewes, 787 ewe hoggs, and 36 rams. The number of lambs born in the spring and alive on the 1st of May last was 1373."

Cause of death.	Sheep and lambs.
Inflammation of bowels	31
Inflammation of lungs	30
Inflammation of udder	1
Died in lambing	14
Abscess, stomach	2
Yellows	2
Heart disease	1
Louping-ill	5
Joint-ill	1
Fluke	2
Found dead	1
Found drowned	1
Found hanged in nets	1
Cripples (lambs under one month old) . .	7
Various causes (lambs under one month old) .	36
Sold for head staggers	28
Sold delicate	4
Total	167

If every farmer kept a register of this kind, not for sheep only, but for all descriptions of farm stock, we venture to say that the annual losses would soon not be half what they now are, and that the result in saving to the nation would amount to many millions of pounds sterling annually.

CHAPTER XXVIII.

DISEASES AND REMEDIES.

Lamb Disorders.—Lambs are subject to very many diseases, some of which are inherited and some acquired. It is, however, not merely necessary to understand that lambs are exceptionally liable to attacks of disease, but that their disorders, if more fatal, are at the same time more easily preventible. However well lambs may be cared for, all of them were not destined to enjoy good health. Still, it cannot be disputed that most of their complaints are due to misfortune rather than fate. As a rule, the younger an animal is, the less heed is given to its wants in the case of sickness, and for that reason many thousands of lambs are annually allowed to perish without any effort being made to save them. When a lamb dies people are apt to say, " It was a bad beast at any rate," and no regrets are ex pressed at the loss. It is true, the lambs which die very often are bad beasts at the time of their death, but perhaps they were not always so; and consequently the loss should be reckoned as if every animal was a good one, which in all probability it would have been had not disease dwindled it into a poor and emaciated condition. At certain seasons of the year the care and management of lambs engage a very large share of attention, and, without confining our remarks strictly to blackfaced sheep, we shall endeavour to describe a few of their more prevalent disorders, with instructions for prevention and remedy.

SCOUR.—Lambs suffer from this complaint at all ages. In very young animals the affection is clearly due to the condition of the ewe's milk, although it may be aggravated by cold or exposure. The evacuations vary in colour, and when inflammation of the bowels ensues sometimes they are mixed with blood, which fact evinces a dangerous condition. The most effective remedy for "milk-scour" is at once to change the food of the dam. After lambing, ewes are very often fed on rich concentrated foods, and in such cases the diet should be restricted to grass or roots only for a few days, which will have the effect of reducing the quality of the milk and rendering it more palatable to the lambs. Purification of the milk may also be hastened by giving the ewe a doze of six ounces of salts in the same weight of water or thin gruel sweetened with molasses. Administer to the lamb one fluid drachm of tincture of rhuburb and a teaspoonful of castor-oil, following this with a dose, two or three times a day, as required, of the following mixture : Prepared chalk, 2 oz. ; catechu (in powder), 1 oz. ; ginger (in powder), 1 oz. ; opium, 1 drachm ; peppermint water, 1 pint. The dose for a lamb under a week old will be one tablespoonful, and should be given in a little cold-flour gruel. For older lambs affected with "grass-scour" give dry food and the same medicinal treatment, only in larger quantities, according to age.

According to Tellor "white-scour" in lambs is attended with much colic, loss of appetite, and rapidly-increasing weakness. In all cases this arises from non-digestion of the ewe's milk. Either the lamb has a weak stomach, or overloads it, or the milk is not of a healthy character. Highly-fed ewes are specially liable to have this disease in their lambs, their milk probably being too rich. In addition to a change of food, he recommends an alkaline laxative to clear the bowels. For this purpose mix together half an ounce of bicarbonate of potash and half an ounce of calcined magnesia. Divide into eight powders, and give one four times a day until the character of evacuations changes. If the weakness be threateningly great, he recommends to beat up two eggs, two ounces of whisky, and one drachm of essence of ginger, in a pint of oatmeal gruel made with milk, and to give a few spoonfuls of it every three hours.

CONSTIPATION.—This is a complaint exactly the reverse of the former, but it is also due to the condition of the milk. A laxative diet should be given to the ewe. Treatment of the lamb may be best effected by giving a tablespoonful of castor-oil and by injections into the rectum. For an injection use warm milk about the temperature of the body, coloured to a light brown with molasses stirred in. Two or three ounces of this are injected with a small syringe. In performing this operation and a moment after, hold the lamb up by the hind legs, so that the fore-feet just touch the ground. If the hardened dung is not discharged with the fluid, repeat. If the animal requires a tonic, give golden sulphur of antimony, 1 to 2 drachms; common salt, 1 drachm; for a dose, once a day.

SORE MOUTHS.—Thrush of the mouth, techically called "Aptha," is an enzootic malady, and resembles to a certain extent epizootic aptha, or foot-and-mouth disease. Enzootic diseases are indigenous to the soil, and arise from an excess of manurial elements in the food, which find their way into the blood. An excess of septic organisms in the system gives rise to sanguineous extravasation, commonly termed "sore mouth," besides being destructive to the constitution. In the animal attacked with "sore mouth," fever is present in the first instance, and shortly a crop of vesicles is observed upon the lips and tongue. The sores sometimes extend to the limbs on account of the lambs

R

rubbing their lips with their feet. "This disease," a correspondent says, "is contagious. In my recollection one lamb had a bad mouth, the dam's teats also got bad, so that it was not allowed to suck, and by stealing its milk from other ewes it affected a large number; and, in the course of a few weeks, thirty lambs had to be taken from their dams and reared on cow's milk, as many of the teats crusted off in milking, and lost the whole nipple. A more deplorable case I never saw in a flock." When an outbreak of this kind occurs the ewe and lamb should be taken from the rest of the flock. The lamb's mouth should be dressed with carbolate of zinc lotion, the strength of ten grains to an ounce of water, and occasional applications made of glycerine or olive oil. It is particularly important to preserve the teats of the ewe, and should the least abrasion be noticeable, dress immediately with a weak solution of blue-stone (sulphate of copper) as a wash. Salves should not be applied to the udder, else the lamb will refuse to suck.

BLINDNESS.—Lambs are frequently affected with sore eyes, a disorder most prevalent in cold wet weather. Very young lambs suffer more severely from this complaint than older ones. It prevents them from following the ewes, and not sucking regularly quickly reduces them in condition, and may prove fatal if not attended to. Blindness is caused by cold in the eyes of such a severe character that acute inflammation is set up, resulting in loss of sight. The best remedy is to keep the little animals sheltered from cold winds, particularly when blowing from the east; and as soon as one shows signs of having weak eyes, bathe them with the following lotion:—New cow's milk, six parts; water, six parts; laudanum, one part; apply with a clean sponge.

LIVER DISEASE.—A disorder far more common than is

generally supposed. It is, perhaps, the most fatal disease to young lambs of all, while its character often remains obscure. It is most prevalent among lambs born early in the season. Lambs born late in spring, when the weather is genial, are never attacked. Shelter is the best preventive, for, without it, the lambs soon get chilled through if the weather be severe, and then it is too late to restore many of them to health. The symptoms of liver disease are fever and pain on pressure of the right side, coldness of the ears and extremities; the bowels, at first loose, become constipated, and the fæces are glazy; the urine scant and very deeply coloured, and the wool is harsh and dry. If an examination be made of the liver, it will be found that the organ appears extremely congested, whilst the other organs present a normal condition. Keep affected animals warm and dry. Give a dose of castor-oil; afterwards give bicarbonate of potash, one scruple; calomel, two grains; opium, two grains, in gruel once a day.

JOINT EVIL.—This disorder is most prevalent in seasons of little sunshine and when the lambs have to lie on cold, wet ground. Cold is the motor of so many disorders that the advantages of providing ample sheltering can hardly be over-estimated. The characteristics of joint evil are stiffness in the limbs, swelled knees and hocks, great lameness, and progression on three legs generally. Every movement evidently causes much pain, and affected animals prefer to lie a good deal, only venturing to follow their dams when forced by sheer hunger. Give the animal daily one and a-half grains of bromide of potassium for ten days, and the following liniment may be applied to the joints and well rubbed in :—Iodine, ½ oz.; glycerine, 2 oz.; mercurial ointment, 2 oz.; olive oil, 6 oz.; mix.

BALL IN THE STOMACH.—In sucking, the lamb is apt to

gather into its mouth any superfluous wool about the udder, and swallow it unconsciously. A depraved appetite, due to acidity of the stomach, causes the lambs to pick up pieces from the ground, and they occasionally pull and eat the wool from their dam's fleece. This craving has also been attributed by some to an irritation of the skin. The wool being indigestible, it eventually forms into a hard ball in the stomach, and kills the lamb. Many lambs are lost in this way every year, and prevention is by no means easy to assign. In former days sheep farmers used to pull off the wool around the udder of their ewes previous to lambing. This practice was termed "udder-locking," but it has generally been abandoned. A few farmers still adhere to the custom, and they speak very favourably of the result. Physicking the lamb is of little avail at a late stage of illness. It has, however, a beneficial effect when given early, and may nullify the craving for foreign substances. Access to salt and maintaining the ewes in a healthy state so as to keep their milk sweet and digestible is, perhaps, the only true remedy for this disorder.

RICKETS.—Every lambing season is marked, in some parts of the country, by the occurrence of what is called rickets or paralysis, causing serious losses among lambs. In some instances lambs are born suffering from what seems to be a general want of vitality. They are incapable of standing, or, indeed, of rising from the ground, although they can move their limbs quite freely while lying on their sides, and frequently struggle ineffectually to get upon their feet; but if by chance they succeed in getting up, or are assisted to the standing position, they stagger for a few moments, their hindquarters give way, and they quickly fall back to the former position.

Lambs which are not attacked until they are a week or two old do not become so entirely helpless as those which are affected at birth; they are capable of rising, and can maintain the standing position, but when they attempt to walk the gait is very peculiar. The hindquarters sway from side to side, and occasionally the hind-legs slip under the body, as though the connection with the trunk were incomplete. No signs of acute febrile disturbance have been observed in any of the cases which have been investigated, except when swellings of the joints occur when associated with severe pain, which is necessarily attended with constitutional excitement. If the sick lambs do not feed, it is generally because they cannot reach the teats of the ewes, and when food is supplied to them by means of a bottle they take it readily. Enough has been ascertained to prove beyond doubt that the so-called rickets or paralysis is a state depending on more than one form of disease; and further, that deficiency of the mineral matter of bone, which is the essential feature of true rickets, has little to do with the condition under consideration.

One cause of the apparent loss of power in the hind-extremities, which is noticed in the disease among lambs, is elongation of the round ligament connecting the head of the thigh-bone with the cavity in the hip-bone in which it is lodged, and in which it should be firmly held by the ligament referred to. The lengthening of the ligament allows the head of the bone to slip out of its cavity at every movement of the limbs, and occasions the staggering and feeble gait which is always present in the affection. It is to be noticed, in reference to the accidents and diseases of the lambing-pen, that while they may be prevented by previous good management, they are generally irremediable when they occur. Most of the victims are in a feeble condition at the time of the attack, and there is very little hope of success, even when the greatest care and attention can be bestowed on the sick animals; but under ordinary circumstances, especially when large numbers of animals are affected, very little time can be given to each case, and the chances of cure are diminished accordingly. The only thing which can be done is to keep ewes under favourable conditions in regard to feeding and shelter during the period of gestation, so as to secure hardy lambs which may be able to resist the severity of an ordinary spring.

RICK-BACKED ANIMALS.—This ailment amongst lambs is attributed to faulty nutrition and similar causes. It is especially liable to occur when ewes have been kept during winter on succulent grass which has lost its goodness, or after a good root season, when the ewes have been fed wholly on this watery food. It may also be expected where the ram has been overworked and in a weakly state at the time of service. Cold, wet, stormy weather likewise favours its appearance. The treatment to be adopted consists in supplementing the mother's milk, or, if necessary, substituting it entirely by cow's milk. Give at the same time half a pint of strong, well-boiled linseed gruel daily, along with "a tonic of six or eight drops of tincture of the chloride of iron."

CURD ON THE STOMACH.—Many lambs die from an accumulation of curd on the stomach. It appears to be unconclusive whether it is poverty or richness of the milk that produces this disorder. It may be partly both. A sudden change from a bare or a poor pasture to a very luxuriant or a very rich pasture is apt to produce it. The lamb in the first instance is insufficiently nourished, and the effect of the sudden change of pasture is to produce an increase in the flow of milk, which proves too much for the lamb in its lowered condition. The probability is, that in nine cases out of ten the evil is due not so much to unwholesome milk as to a weak digestion in the young lamb, which ultimately dies of distension of the stomach. It may be said that a sucking animal is not likely to take more milk than it can digest. Experience shows, however, that many calves as well as lambs die from this cause every year. Many calves are killed by overfeeding them during the first few days of their life, and particularly when they are gorged with skim-milk. The digestive organs of the young animal are not prepared for this all at once, and the lamb or calf succumbs. It is much the same when an older lamb, with a weakened constitution and digestion, is allowed a sudden increase in the quantity of milk. Were the lamb able to digest this milk, the effect of the changed keep would probably show itself in the form of low fever or scour. The cause and remedy of both, in the practice of those who, in all such cases, find the benefit of shifting the ewes and lambs. An alkaline powder, of the kind recommended for " white scour," which is only a more acute form of this disorder, should also be administered.

Sturdy.—This is a parasitic disease which develops in the brain. It is common to every breed of sheep, although it only attacks those of a tender age. It is very rarely found in sheep over a year old, from seven to eleven months being the most virulent period of the disorder. The cause is due to the ova of tape-worms scattered by dogs on the pastures. These find their way into the stomach of the sheep, and carried, it is supposed, by the circulation of the blood, they develop in the brain of the sheep. The ova, necessarily very minute to commence with, form into a cyst or bladder, and this, by its gradual formation, interferes with the structure and functions of the organ, giving rise to peculiar movements, from which the above term, with others in common use, such as " gid " and " turn-sick," have been derived.

When a sheep is affected with sturdy, the seat of the disease may be detected by the animal's movements, as well as by pressing at different parts of the skull. As a rule, a sheep suffering from sturdy, walks, runs, or totters in a circular course, or in a straight line, with elevated head, whenever the bladder worm or hydatid is situated in the middle of the brain; and it hesitates in moving, falls, and even rolls over, when the hydatid is at the back part of the skull, or so-called crown of the head, and beneath that part which is between the horns in horned sheep. When the animal turns constantly to the right or to the left, the hydatid is situated towards the forehead on the right or left side, opposite to that to which the animal turns. As a general rule, if the sturdy be at all severe, the position of the bladder is detected by feeling for that portion of the skull which may be soft and thin, owing to absorption from pressure of the sac beneath.

The cure for sturdy is accomplished by tapping the bladder and drawing off the water, which varies in quantity from a tea-spoonful to a table-spoonful. This operation is best performed by a veterinary surgeon, but any intelligent shepherd may soon become an expert at the work, provided he gets sufficient practice. Most farmers provide themselves with the instruments required, and therefore are in a manner independent of professional assistance. Those who may be in want of a set of such instruments, and not knowing where to get them, may be told that the Messrs. Hilliard, Nicolson Street, Edinburgh, furnish a very serviceable case at 16s. 6d., which will be found complete in every respect.

Having first bound the sheep's legs, and determined the position of the parasitical sac, the wool around is clipped, and if the bone over it be thick, the borer, provided with a nut which may be screwed up or down, is employed, and the nut is placed at such a distance from the point of the instrument as to allow of perforation of the skull and bladder only, without going deeper and injuring the sheep. If the skull be very thin, the trocar only is used, and steadily pushed in; the canula or tube being pressed, and the stylet withdrawn so soon as the puncture is made, in order to allow the escape of fluid through the canula. Whether the trocar be used alone, or a passage prepared for it by the borer, as the fluid is drawn off by the canula, the bladder appears at the opening; it is generally essential, but not always advisable, to use the syringe to draw the liquid out through the canula, and having done this, the forceps are used to lay hold of the portion of the bladder which presents itself at the opening in the skull, when it may be withdrawn. The operation, however, may be quite successful without the removal of the bladder, and if it cannot be conveniently extracted, it is better not to cause unnecessary irritation in search of it. The wound should then be wiped clean, and bound over with a linen cloth. The danger of the operation consists in producing inflammation of the brain, which, should such subsequently arise, proves rapidly fatal. If the sheep are in a condition fit to kill or go to the butcher, it is often a more preferable course to adopt than to attempt to cure them. About 70 per cent., however, may be cured by the process described, but they are never afterwards so valuable sheep as if they had never suffered from the complaint. Farmers may largely prevent this disease by keeping all the dogs employed on the farm free of worms, and by carefully destroying infected heads of dead sheep.

Braxy.—This is perhaps the most prevalent as well as the most fatal disorder known among hill sheep. It is common to nearly every district in Scotland, and also in several parts of England and Wales. It has never been traced as originating from any particular geological formation, yet it is vastly more general in certain localities than it is in others. Braxy or "sickness," as it is sometimes called, is a disease confined for the most part to the younger members of the flock, hoggs or lambs during their first winter only being subject to its attack. The malady almost invariably occurs in the autumn months of the year, at which season it is most fatal, although in spring occasional instances of its appearance are not unknown. The cause of Braxy has, on frequent occasions, been a subject of much controversy among both practical flockmasters and scientific men. It is, however, now generally admitted that a sudden change in the food from fresh succulent grass to dry or frozen herbage is the chief cause of the disease. This theory is supported by several noticeable features. In the first place, Braxy never appears during the summer season while the grass is soft, nor is it associated with any kind of green food whatever. On the other hand, as soon as the natural herbage on the hills begins to lose its freshness, and becomes withered or frozen, the disease is certain to appear. These facts are sufficiently well known to all shepherds. Indeed, so nicely do some of them calculate upon the freshness of the pasture, that after going out in the morning and observing the rhyme and frost upon the ground, return to the house to put on an old hat and coat in the expectation of having to carry home dead sheep. And their fears were seldom groundless, as on going to the sheep as many as five or six have been found dead or dying in a single morning.

The rapidly fatal character of the disease renders investigation extremely difficult, and scarcely leaves any room to attempt a remedy. Sheep that appear perfectly well and healthy at night will be found dead in the morning.

Professor Williams, and some other members of the Veterinary profession, consider the disease analogous to, if not identical with, Anthrax. All its symptoms—and it has several—are inflammatory, whether it be seated in the bowels or in the flesh. The progress of the inflammation is exceedingly rapid, but in the later stages, when mortification sets in, the pain ceases, and the sheep, though apparently improving, usually dies very suddenly. The first indications of illness are restlessness and frequent changing of position, with a dull sickly appearance. The animal affected, if grazing along with others, may be observed to cease feeding, occasionally lying down and rising up, and eventually separating from its neighbours. The head hangs down, and there is sometimes a rapid lifting of the hind feet, and a crunching noise with the teeth indicative of pain. The bowels are costive, the paunch begins to swell, and the back is slightly arched. When the pain becomes more acute, the patient ceases to rise, and lays its head on the ground in silent agony. At death, the mouth and nostrils are usually filled with froth mixed with blood, but the most noticeable feature connected with this disease is the intolerable stench which accompanies it in every case. As soon as the carcass is opened, the foul gas which proceeds from the viscera cannot be mistaken if once experienced.

On dissection, it will also be observed that the inflammation is either confined to the bowels or solely to the fleshy parts of the body. This is the most mysterious circumstance pertaining to the disease, and one which has never been satisfactorily explained. How it happens that the inflammation should appear in the flesh of one animal, and be confined solely to the intestines of another, is a deeper point in connection with the disorder than has yet been reached. Still, there is no doubt whatever as to their being one and the same ailment, as the very smell itself, and other similar symptoms in both cases, sufficiently prove.

But besides this difference in the condition of the dead animal, braxy has other symptoms peculiar to itself. Sometimes the animal attacked is seized with a form of the disease usually termed water braxy, and when a sheep dies affected in this way, it is found to be filled with a bloody kind of water. Where so much water can possibly come from, seems impossible to tell, but it is there nevertheless. In such cases the gall and bladder are quite empty, and although the water in the sheep is outside the entrails and all natural channels, the supposition that the bladder may have burst is contradicted by the fact that it can be filled with air, the same as one from a healthy animal. While treatment may in some cases prove successful, braxy is usually considered incurable.

When taken at an early stage a strong dose of Epsom salts with some ground ginger has been found useful. A correspondent of the *Irish Farmers' Gazette*, who states that he has succeeded in restoring several sheep attacked with braxy to perfect health, places his dependence chiefly in linseed oil given several times a day. He recommends 2 oz. linseed oil and 3 grains powdered opium in a little linseed tea. Repeat the opium the following day along with two scruples of ginger, and if the bowels continue costive the oil may again be administered.

Bleeding in the early stages can be practised with a certain amount of success, but after the disease has reached an advanced stage the blood cannot be drawn by any ordinary method. This is another remarkable feature of the malady, showing that the circulation of the blood is wholly deranged, and like all other anthrax diseases impossible to cure. Prevention is the most satisfactory remedy for braxy, only in the case of mountain sheep it is sometimes difficult to apply with advantage. It is well known that to remove the hoggs to turnips is a sure preventive, but this system of feeding is not adapted for hill sheep. If the disease should be so prevalent as to render removal necessary, then grazing of some kind is to be preferred to turnips. The expense of taking the hoggs to wintering is of course a heavy item, but it is not so great as it appears at first sight. When the hoggs are removed in winter altogether a larger stock can be maintained upon the farm, which in some measure repays the extra cost of wintering. However, a good deal may be done to avert braxy by attending to the proper drainage of the land, and also by judicious herding.

Could some means be devised that would ensure the growth of hill pastures all through the winter, there would be no more braxy. Another efficacious remedy, if it could be carried out, would be to grow some winter herbs or plants on the hill ground, fencing them off until such time as the disease was apprehended, and then giving the hoggs a run over them daily. These methods may seem impracticable as they are in the meantime; but if some clever farmer or seedsman could produce a plant suitable for this purpose the possibility of preventing braxy would be no longer an unsolved problem but a grand reality.

A very extraordinary system of treatment for anthrax and braxy has recently been communicated to the Highland Agricultural Society by Mr. Buchannan Arbuthnott, Fordoun, which we reproduce, as since it was published, corroborative testimony has been given of its beneficial effects on sheep by others in different parts of the country. It is as follows :—

"My remedy, and I think I may fairly claim the discovery, is as simple as it is effective. The treatment consists of a species of inoculation by which the young animal is freed from the liability to the disease. A preparation of turpentine and garlic is inserted under the skin in each quarter, and so quickly does it permeate the system, that within a few minutes of its application the smell of the ingredients is perceived in the breath of the animal. The art of the operation is in the way of performing it, so that the remedy shall remain inside the skin, and not come out until the animal is killed and skinned, when it will be found as fresh and active as when first put in. I have long been under the impression that what is called braxy in young sheep is the same as quarter-ill in cattle, and last September I inoculated one hundred ewe lambs for Mr. John Campbell Baillie, Farfside, Glenesk, with the result that not one out of the hundred has died : they were sent to the low country for the winter along with other three hundred, and out of that number twenty-one have died, the whole flock getting the same treatment.   The expense is so little that it is hardly worth writing about.   Sixpence would more than do twenty cattle, and half-a-crown one hundred sheep.

"I seek no personal emolument, being content to serve my fellow-man in the matter, so far as I may, by helping him to eradicate the evil. Get from a seedsman ½ lb. finest . French garlic bulbs, remove the skin, and break every section of the bulb from one another ; these all again have to be skinned ; after which take a jar or wide-necked bottle, put in the garlic as prepared (that is the soft juicy parts), and then put turpentine into the bottle sufficient to cover the garlic ; cork it up for twenty-four hours,—it will then be ready for use.

"It will keep as long as you like.   Treatment of the animals :—They are inoculated on all the four quarters, on the flat of the fore and hind legs, on the thigh in the most convenient place.   Take the skin between the finger and the thumb of the left hand, make sure and draw it well from the flesh, make a horizontal cut with a sharp knife sufficient to admit your little finger, which insert to remove the skin from the flesh in the direction of the animal's foot, put the finger as far down as you can get it.   This makes a hole exactly like a pocket.   Put in one of the sections of the garlic, and leave it there."

Louping-ill.—This is a fatal disease prevailing to a great extent among hill sheep throughout Scotland. It is not confined to any particular geological formation, although it seems to be most destructive in the Silurian districts.   It invariably occurs from about the beginning of May to the middle of June, sooner or later, according to the character of the season and the forwardness of the herbage. In Skye, however, there are said to be two annual outbreaks of the disease—one in early summer and another in autumn—both of which continue for a regular period of from six to eights weeks. The autumn complaint is not known in any other locality.

Of the real cause of the disease little is known of a definite nature. Formerly it was supposed to have been caused by cold east winds, or from the sheep partaking too freely of withered grass, an article very abundant on hill pastures. These theories, however, are untenable. The disease is quite as prevalent on the sunny sides of the hills as those which feel the full severity of the north-east winds; and as to withered grass it is plentiful in many districts where louping-ill was never known. Still there is no doubt that a change in the weather from warm to cold in some way or other predisposes to the disease and causes many animals to be affected that would otherwise most probably have escaped without serious consequences. In certain districts it is questionable if there are any of the sheep that do not at one time or other pass through the infection, but have it so slightly as to be unobserved. And when exciting causes are present, such as a change in the atmosphere would occasion on a debilitated constitution, the disease is revealed. Another reason which proves many sheep to be infected that do not appear so, is often seen when they are scared by a dog by several of the flock becoming suddenly ill. Careful shepherds keep up their dogs, and treat their sheep as quietly as possible whilst the danger exists.

In Roxburghshire, where great losses are sustained every year from louping-ill, the Teviotdale Farmers' Club, being suspicious of another cause, engaged Mr. Brotherstone of Kelso to examine the flora of the district in order to ascertain if any poisonous plant prevailed on the unhealthy tracts which would be likely to produce the effect. In an able report, Mr. Brotherstone stated that he could find none, but that in the worst places ergots were very prevalent on the grasses when in seed. As they do not come into seed, however, till the season when the louping-ill is over, he thought they might remain in or among the withered grass, so as to be eaten with it in the spring. On receiving the report the committee engaged Mr. Webster, veterinary surgeon, to try the effect of ergots on a few sheep directly administered to them, and also to dissect some of the lambs which had died of the disease, and ascertain the seat of the malady. In May 1882 four one-year-old sheep were accordingly obtained from a district where louping-ill is unknown, and were fed for fully a month on ergotised grasses, carefully collected by Mr. Brotherstone in an infected district. Six more ewes and lambs were added to the number and fed in a similar manner, but no effect was induced further than a slight purging.

The late Professor James Elliot was a member of the above committee, and as the result of his observations and enquiries amongst farmers and shepherds, wrote a very plausible theory of the disorder, which, like some others before him, he maintained was caused by the bite of the tick (*Ixodes Ricinus*), an insect very common in the infected localities.

The efforts of the Hawick Farmers' Club being unsuccessful in solving the mystery, Professor Williams and Dr. Aitken, veterinary and chemical advisers of the Highland and Agricultural Society, were then called in. These gentlemen made several visits to infected districts to trace, if possible, the source of connection between the tick and the disease. A great number of cases were dissected, and in all the subjects Professor Williams detected a jelly-like formation within the spinal canal, extending in a more or less uniform layer from one end of the canal to the other, but rarely extending within the cranium. The jelly, when examined microscopically, was found to contain an organism, the *Bacillum choræ ovis* which he considered sufficient to cause the disease by irritation of the nervous system through the spinal cord, in the fluids of which the organism finds a suitable nidus for development. Professor Williams further discovered the same rod-like bacteria in the tick which had been obtained from diseased sheep. This proved that the organism was in the tick, but whether the tick received it from the sheep or the sheep from the tick was still unsettled. However, ticks from healthy sheep were obtained, and after repeated trials the same organisms were found to be developed.

It has not yet been ascertained whether the disease can be inoculated from the tick to the sheep, or from one sheep to another, and until such experiments have been made, Professor Williams's diagnosis remains incomplete; but there is every reason to believe ticks are the agents through which the disease is transmitted. Whether the sheep are affected internally from what they eat, or are externally inoculated by the bite of the tick, is as yet a mystery. But however closely ticks may be allied with the disease, it is hardly possible that they could produce such effects either by irritation or extraction of blood from the sheep. And this theory is corroborated by the fact that in many instances few parasites are to be found upon diseased sheep, while many healthy ones are swarming with them, and young lambs have been known to die of the disease before the age at which they begin to eat grass. Still it is not beyond possibility for young lambs to receive it in their mothers' milk, and it is rather a pity no analysis, as far as we are aware, has been made of a ewe's milk from which a lamb has died. Were a ewe to be affected while pregnant,

S

and bring forth her lamb in that condition, the offspring would also unquestionably inherit the disorder. It might be mildly, yet in certain cases, as is sometimes witnessed, sufficiently severe to cause the death of the lamb. In such cases the lamb is too young to eat grass, and neither are there any ticks to be found, so that we have here a hereditary cause, which seems to be entirely overlooked by the investigators in accounting for the death of lambs.

The disease being most fatal to lambs, however, at the age at which they commence to eat grass, it would appear that the poison is present in the food, and it need scarcely be stated that it would affect lambs more injuriously than old sheep. An analysis of the herbage having revealed nothing deleterious in any of the grasses found, it would seem that if ticks are really the cause they must be eaten by the sheep along with their food, and act as a direct poison in the stomach. But as ticks are found in many localities free from louping-ill, it is concluded that they are conveyors of a poison existing in the soil. Yet as the disease is not confined to any particular formation or soil, and can be transferred by infected sheep from one district to another where it previously was unknown, the supposition of the soil being the original source of the organism falls to the ground.

The tick is only found on land where a quantity of rough and withered grasses are allowed to remain. Where the pastures have been improved by closer eating with cattle, draining, liming, or other means, and no previous years' growth to shelter the parasites, it invariably disappears, and the louping-ill along with it. At present this is the only known remedy, but unfortunately it is impracticable on most hill ground.

Pining or Anæmia.—Pining is perhaps not so much a disease as the result of malnutrition. It is, however, common to hill sheep in nearly every district. In the spring months of the year the leanest of the flock are usually carried off with what the shepherds term "poverty," "pining," or "hunger-rot," all of which refer to this disorder. Scientific men have termed it "anæmia," or bloodlessness, which means the same thing. Lambs or hoggs are more subject to this disorder than old sheep, for reasons which will hereafter be apparent. In some the symptoms of "pining" are thriftless-ness, losing their wool, and a "papery skin;" others cough and waste; the usual remedies for bronchial filaria do not benefit them; and *post-mortem* examinations fail to disclose worms in the air-passages. Others are dropsical about the throat or belly, and the shepherds report death from "white water" in the abdominal cavity.

Flukes are looked for, but in many cases are not found, or are not in numbers sufficient to account for death. Most patients are relaxed in their bowels, feverish, very thirsty, with irregular appetite, flesh gradually melting away, sunken eyes, and are usually carried off by dysenteric purging. All these conditions proceed from the same cause—poverty of blood.

During winter, or during the wet weather of autumn, the lambs are liable to receive a check. Pasture may contain abundance of coarse long grass, perhaps soaking wet, as is often the case, and lambs do not willingly tackle such fare. They will often rather famish than feed on it. Food, un-less it is eaten, digested, and assimilated, can do the crea-ture no good. This truism is much overlooked in the management of young stock. Although the casual observer may not observe any special falling off in the health or con-dition, the blood of anæmic animals was certainly starved at some period. From want of suitable nourishment it underwent degenerative changes. The blood-globules be-come reduced in number, and contain less colouring matter; the faulty state of the nutrient fluid, the blood, is propa-gated throughout the body; the pale blood percolates through the thin walls of the vessels, producing dropsical effusions; the mucous membranes and skin are paler and softer from inadequate nourishment; the wool is dry and broken; diarrhœa appears.

After a time the muscles shrink, and are sometimes paralysed. Anæmia is more widespread and fatal in seasons and localities where young stock have been in unfavourable conditions as to wet, exposure, or indifferent food. It pre-eminently attacks young and growing stock before their tissues are consolidated and their full strength attained.

Mischief done weeks or months ago is now evolving; the malnutrition of most organs in the body is not easily arrested. Careful nursing will do more than medicine. Proper food is more wanted than physic. The stomach needs relief rather than more work, hence the food should be lighter and given more frequently in smaller quantities at a time. But when all is said and done, it is one of the hardest things a man ever undertook to bring back an anæmic sheep to normal health and strength. The disease is very slow in coming on, and it is still slower, if anything, in going off. When one has done all he can he is pretty sure to lose a few sheep out of an anæmic flock, and often these very sheep are the ones he considered the nearest cured. The leanest of the flock should be drawn out early in spring and taken into more convenient quarters where they can be properly attended to.

It is a great mistake to leave them on the hills to drop off one by one without attempting to save them. If they can be turned on to a little fresh pasture for a short time daily it will be an advantage, but their chief fare must consist of a little of the best hay and a few crushed oats and bruised linseed cake. Along with such concentrated dry food should be given daily about thirty grains of powdered sulphate of iron.

As a serviceable vegetable tonic a drachm of powdered

gentian is administered with three or four ounces of old ale or of whisky and water.

A solution made by boiling oak-bark in water often proves an effectual astringent for arresting the wasteful diarrhœa.

Lung Disease.—Lung disease is a common ailment amongst sheep, and it has been said that the disease is not only contagious, but probably also hereditary. We have very good veterinary authority, however, for saying that there is no form of lung disease in sheep the character, course, or history of which warrants us in attributing to it infectious or contagious properties, and it is equally certain that the disease in question is not hereditary. A careful examination of such cases will invariably show that the affection belongs to the common forms of lung disease (*broncho pneumonia*).

Sheep are particularly liable to coughs and cold, their habitual residence on the land, where they are exposed to all kinds of weather and changes of temperature, being sufficient to account for this tendency. Very little notice is taken of ordinary catarrh amongst sheep; and if by chance any losses occur from the extension of the irritation to the bronchial tubes, and the subsequent development of congestion or inflammation of the lungs, they are generally ascribed to other causes. Some of the present races of sheep are not characterised by hardiness of constitution, and when the weather is severe, ordinary catarrh very often assumes a malignant form. There is little or no room for doubt that the disease of the lungs from which sheep have suffered severely in some parts of the country during recent years has its origin in an ordinary cold from exposure. The disease prevails as an enzootic during severe seasons, where the sheep are in a debilitated condition and shelter is neglected. The primary cause can also often be traced to sudden chills of the system. The sheep most liable to succumb to this disease are old ewes and weakly hoggets.

Some of the latest reported cases in Scotland are amongst hill hoggets that have been sent to wintering on turnips.

The preventive and remedial systems of treatment will be in all essential details the same. Shelter and good food, in substitution for exposure and low diet—low, that is, as to nutritive value, and not merely deficient in quantity—are, of course, the first necessities. But medical remedies, in addition, are useful. The judicious use of saline alteratives and mineral tonics, under competent directions, will seldom fail to bring the patients back to health.

Rheumatism.—Rheumatism sometimes attacks lambs. The limbs, or some of them, become stiff, causing a difficulty and awkwardness in motion. There are cramps in the neck, and the animal manifests an inclination to remain quiet and listless. The bowels soon become constipated. One writer states that if ewes are fed during the last months of pregnancy on mouldy food of any kind, it may cause rheumatism in the lamb. We give this statement for what it is worth. Certainly the animal should not be fed on mouldy food while in that condition, whether rheumatism result or not. Such food would readily produce abortion, if nothing else, but it has also a bad effect on the milk. In the first place, provide shelter, and see that the ewe has proper food. Give the lamb the following at the commencement of the disease :—Powdered sulphuretted antimony, five parts, and fresh butter, one part. Mix, and then administer a quantity the size of a hazel nut three times a day.

Strongles.—Thread-worms, as they are popularly called, inhabit various parts of the organisms of the higher animals, and the mischief they do is in proportion to the extent of the obstruction which they occasion to the function of the part. In the intestines, for example, they seem to cause little disturbance; in fact, in many cases their presence is not known until an accidental inspection after death reveals them. As in the case of flukes, the administration of the eggs of the mature worms is not followed by any positive results, and the direct introduction of them into the air passages has proved equally abortive so far as parasitic infection is concerned. It may therefore be assumed that the eggs which are expelled from infested animals undergo certain changes on the pastures where the sheep are grazing; and the result is that lands once grazed over by animals which are infested with strongles in the intestines or bronchial tubes become contaminated with the larval form of the worms, and other animals following the infested ones over the same ground some time afterwards become also infested; and this fact is so well known that flockmasters, as a rule, avoid letting lambs follow ewes over clovers, their experience being that the young animals suffer seriously, while the older sheep, which are the sources of the parasitic infection, do not indicate the presence of the worms by any outward sign, owing, probably, to their superior powers of resistance to the invasion. The eggs of strongles are not quite so much dependent for their development on the character of the pasture, as are the eggs of the fluke; but there is no doubt that a moist locality is most favourable to them, and in fact a hot, dry season is fatal to their further progress. Pastures which lie low, or which are subject to occasional overflows, are the situations where strongles flourish; and land once infected retains its character permanently, unless some steps are taken to eradicate the germs of the parasite, and these steps must include the entire disuse of the pasture for a season or two, and the use of a top dressing of lime of say ten hundredweight to the acre.

It is almost needless to say that precautions are not as a rule taken at all, and things are allowed to go on in accordance with the law of accident. Scientific men, in dealing with the problem of the restoration of contaminated lands to a healthy condition, are met by the difficulty of accounting for the advance of the larval forms of the strongle towards maturity, and this difficulty they have not yet been able to solve. The symptoms which indicate the existence of the lung-worm in lambs are very well known. A short dry cough or "husk" is the most prominent sign; but as it occurs in the early stage of the disease, being the first result of the irritation produced in the bronchial membrane by the worms, it often escapes notice. Very soon the in-fested animals begin to fall away in condition, the cough becomes more frequent, and some of the weakest lambs die.

The preventive remedy for this, as also the fluke-rot, must be cheap, effective, easily applied, and within easy reach. There are two such remedies within the reach of all, viz., lime and salt, both in different ways correctives and improvers of the soil and herbage, and also beneficial to animal health. Farmyard manure and some artificial manures would tend to foster the production of low forms of life, which would be obviated by the addition or substi-tution of lime or salt.

In using lime to drinking pools there is no danger if ordinary precaution be used, whilst a small quantity in hot weather is highly beneficial. It would require some further experience to estimate the quantity to be used which will effectually destroy the ova and young worms without making the water too brackish to drink. This may be judged from the solubility of lime in water. In summer, 100 gallons of water dissolves about 1 lb. of lime; but as this would be too strong for the animals to be inclined to drink, about half or two-thirds the quantity of lime may be used and repeated daily in two or three days, or an interval of a week or two, according as to whether it be a running stream or supplied by a neighbouring spring, or whether it be a half stagnant or stagnant pool. In either case it should be well puddled.

Fluke, or Liver Rot.—This is a parasitic disease which prevails only on swampy or undrained pastures. Fortunately, it is comparatively little known in Scotland, but cases of it are nevertheless to be met with on almost every farm. Stagnant water, however caused, is necessary for the development of the fluke. Excessive floods, especially in summer and autumn, the supersaturation of the soil, undrained or imperfectly drained land, the effects of a series of wet seasons, rendering strong land, even when drained, so impervious that rain water remains on the surface—all these are causes. That the disease has prevailed on land hitherto considered sound, arises from the eggs voided by diseased animals having found a suitable hatching condition, owing to excessive moisture, which in ordinary circumstances would not be possible.

In the earlier stages of the disease we get no external indication of its presence; in fact, as many a farmer knows by painful experience, the sheep in the earlier stages of the disease have just the opposite appearance of disease, the animal having a tendency to fatten, especially if on a rich watery pasture. It is not until the disease has undermined the general health of the animal that it shows unmistakable evidence of its presence. The animal then becomes emaciated, listless, and stands or lies about the field by itself, the eye becomes dull, and an examination of the small veins in the corner of the eye will show them to be filled with a yellowish fluid, instead of blood; a "poke" forms underneath the jaw. In this state the animal gradually gets weaker, till it at last dies from sheer exhaustion. Where the cause of attack is present in unlimited force, every individual in the flock may perish, and the losses which have been sustained by the fluke disease are sometimes fearful to contemplate. Not only has individual farmers suffered from it, but whole districts have been decimated of their sheep stock.

We can only give a condensed account of the development of the devastating pest. The egg of the liver fluke, *Fasciola hepatica*, is a smooth, transparent, yellowish brown, oval body of microscopic size, its length varying between four and six one-thousandths of an inch, and its breadth from three to four one-thousandths of an inch. A single fluke is capable of producing several hundred thousands of such eggs, so that a sheep whose liver, bile ducts, and gall bladder are infested by flukes, would in a few months pass out in its excreta a most prodigious number. The eggs are not hatched while in the body of the sheep, but in the water or moisture on a pasture. Before emerging from the egg, the embryo furnishes itself with a covering of delicate filaments, called cilia, by the waving to and fro of which it is enabled subsequently to propel itself through water. The ciliated swimming embryo has the shape of an elongated cone, with a rounded apex, and is about a two-hundredth of an inch long. The average duration of the life of the embryo in water is only about eight hours. The question now arises, what becomes of this short-lived embryo, and how is it connected with the adult fluke, as found in the bile ducts of the sheep's liver? It has been ascertained that its tendency is obviously to bore its way into some object, and from the analogy it presents to other closely-allied organisms, whose life-history have been fully observed, it is concluded that it would choose for this purpose the body of some mollusc, as a snail or slug, and after undergoing modification in the host, which in turn gains access to a suitable vertebrate host, as the body of a sheep, deer, or rabbit, it would develop into an adult fluke.

Assuming, then, as there is the strongest reason for doing, that there is an intermediate molluscan bearer, it is necessary to inquire, In what form does the fluke enter the sheep, and in what way? Does it enter with the food, or with the drink, or by some other method? Now, it is necessary for the cercariæ of the liver fluke, like those of its nearest allies, to become encysted—*i.e.*, enclosed in a sac—a condition in which the organisms become motionless, and invest themselves completely and individually with a cyst, and this may happen either in the body of the molluscan host, or much more frequently after the cercariæ have been voided in water, or suitable hatching ground. Judging from analogy with other flukes, it is most probable that sheep pick up the young river flukes, whilst in the encysted state, with their food. They are not likely to be taken up with the water drunk by the sheep, for free cercariæ would perish. Moreover, rabbits are exceedingly liable to flukes— as many as fifty have been found in the liver of a single rabbit—and sportsmen deny that they ever drink. It has also been conjectured that sheep eat slugs, but such is not the case. The more probable theory is, that the cercariæ encyst themselves at the moist roots of plants, and are taken up when the sheep is grazing. This view has much to recommend it, and in accordance with it is the opinion of farmers that the fluke is especially taken up when the sheep graze closely, and that the better the biter is, the less are the chances of escape. Of the molluscs which would support the intermediate stage, or cercarian form, are the common grey slug, *Limax agrestis*, and the black slug, *Arion ater*.

Liver rot being known to exist in the flock, the question of treatment naturally arises, and, says a good authority, there are several points which require consideration. First, it is well to be quite certain whether or not the sheep became infected on the ground where the disease was discovered. A mistake in this direction may be serious, because it is most important that the animals should be removed from the place where the larvæ of the fluke are to be found. It will often be ascertained on inquiry that the sheep have been recently purchased, or that they have been feeding in

other pastures ; and it may be the case that the ground on
which the sheep are at the time of the investigation is
perfectly healthy land.   When it appears on inquiry that
the affection has originated in the pastures where the animals
are still feeding, it becomes a mere common sense proceeding to move them to dry ground.   If from any cause this
essential step cannot be taken, treatment is not likely to be
of much use.   It has been proved over and over again that
rotten sheep have a fair chance of recovery if they are taken,
in the early stage of the disease, to a bare, dry pasture,
where they will be required to hunt for their living, whereas
if they continue on the wet ground, even if they do not take
up more larval forms of flukes, which most likely they will
do, there is little opportunity for the restoration of the tone
of the digestive organs, and the majority of the sheep will
die of debility—many of them after the flukes have all been
expelled.   Undoubtedly the successful treatment of rot
necessitates the placing of the infected animals in a favourable position, and supplying them with dry, nutritious food,
with the addition of salt, both in the form of rock-salt—
which may be scattered about their feeding ground—and
bay salt, which may be mixed with the trough food, so that,
as nearly as may be, every sheep may get about one ounce
per day.

Various specifics for the cure of "rot" have been adver-
tised here and on the Continent, and some of them appeared
to exercise a beneficial influence on the animals to which
they were given.   Generally the best nostrum contained
preparations of iron, combined with vegetable tonics and
stimulants, and would therefore act by improving the tone
of the digestive organs.   No treatment has, however, proved
more effectual than the free use of common salt with good
dry food, and the removal of the animals from the position
where the disease finds favourable conditions for its full
development.

Jaundice or Yellows.—Jaundice is due to functional derangement of the liver, resulting in an imperfect secretion of bile. When the digestive organs are in perfect order the liver supplies sufficient secretion to act in conjunction with the fluid from the pancreas (sweetbread) in converting the mass of food which passes along the first intestine into the chylous mass, and in this way completing the digestive process. In jaundice it appears that the colouring matter of the bile is absorbed and carried along with the alimentary matters into the circulation; hence the universal tinge of yellow which is invariably apparent in the visible mucous membranes of animals affected with the disease, and also the entire surface when the skin is light in colour. Evidently there is something in the pathology of the malady which is not clearly understood. Organic or functional disease of the liver, associated with diminished secretion of bile, will not cause it. Constipation and an altered secretion of urine often result from this cause without any symptom of jaundice being present.

But in the case of there being an obstruction to the escape of bile after it has been secreted, more or less yellowness of the mucous membrane is the consequence. The symptoms of jaundice are sufficiently marked to be recognised by the unprofessional observer, irrespective of the indications of deranged digestion, such as loss of appetite, constipation, absence of colour in the defections, debility, and emaciation—the yellowness of the skin and membrane is one unmistakable sign of the affection. The treatment is usually tentative in its character. Sulphate of magnesia (Epsom salts) is the best laxative in cases of jaundice, and it may be given in combination with extract of taraxacum, which acts on the liver. Tonics will not be required until some impression has been made on the disease; but the combination of a mild stimulant or carminative with the medicine is desirable from the commencement of the treatment. A draught composed of Epsom salts, 4 oz. ; extract of taraxacum, ½ oz. ; powdered ginger, 1 oz. ; and gruel, 1 quart, should be given once or twice a day according to the urgency of the symptoms, until the bowels are acted upon, when the quantity may be diminished, or the medicine given less often, so as to avoid excessive purgation. Compounds of iodine have proved effective in the later stages of the disease ; and as there is always a deficiency of red corpuscles in the blood when debility is present, a combination of iodine with iron furnishes at once a remedy for the impoverished blood and the inactive condition of the gland. In the matter of diet very little discrimination need be used, as the appetite is capricious, and there is not much danger of the patient taking its food too freely.

Garget.—Garget in ewes is a common complaint, both at the time of lambing and weaning, and may be known by the swelling and heat of the udder. Veterinarians would, doubtless, draw a distinction between the two causes, and probably designate them as different complaints. But practically they are disorders of the same nature, and may be treated alike. Sometimes black spots appear on the udder, which break and make very stubborn sores, ending in the destruction of the animal. Fever and lameness are also present. Treatment consists in feeding the ewe moderately, avoiding heating foods, such as corn. Foment the udder, and give internally the following :—Oil of turpentine, half ounce ; sulphate of magnesia, four ounces ; powdered ginger, one ounce. Draw the milk from the udder as frequently as possible, and when these measures are promptly adopted, before the disorder is firmly established, the cure is speedy and certain. Otherwise there is great danger of ruining the udder, and, perhaps, the life of the ewe.

Milk Fever.—Milk fever in ewes is not very common, but it does happen, nevertheless, and not unfrequently proves fatal without the cause being detected. The first symptoms are weakness, dulness, unsteady gait, loss of appetite, and a twitching of the ears and hind legs. It is most common when the ewes are in high condition. The time of attack is usually a day or two before yeaning time. Give the following dose :—Nitrate of potash, one drachm ; sulphate of magnesia, three ounces ; molasses, three ounces. This may be given in warm linseed gruel. This will open the bowels, or if it should not, in about ten hours repeat the dose. When the bowels have been relieved, give twice a day the above dose, with the exception of the magnesia, so long as the fever continues. Another remedy is to give five drops of aconite and nux vomica, alternately, every three hours. After the fever has subsided, give nourishing foods and tonics, such as charcoal powders.

Scab.—Before dipping became known in the practice of sheep farming, very great damage was often caused to mountain flocks by the parasitic disease termed scab. It is now, however, seldom met with. The acarus, or scab insect, belongs to the very minute order *Acarina*, of the class " Arachnida."

It is the same species of parasite as that which causes mange in the horse or dog, and, indeed, infests every species of the animal kingdom. These mites are no larger than the point of a pin, yet they multiply with amazing rapidity, and work the most disastrous results when allowed to settle on a flock. They burrow under the skin and live on serosity, the effusion of which is produced by the irritation which they excite. Any animal may be the bearer of contagion between other two ; but it is essential for the development of a real scabies on any animal that the insect should be proper to that animal. The symptoms of scab in the sheep are easily detected. The animal attacked is seen to bite with its teeth the part affected, and the wool, from being loosened at the roots, begins to fall out. The sheep is in great misery from the itch caused by the myriads of mites irritating its body, and will betake itself to such places as hags and posts, where it will rub nearly the whole fleece from its back. The constant irritation creates uneasiness, preventing the sheep from feeding or resting; and, as a consequence, it rapidly wastes in condition. On the part where the wool has been destroyed the skin assumes a roughish, blistered appearance of a greenish tint, which can hardly be mistaken by any experienced shepherd.

While scab is a most contagious infection, it is easily curable, even in its most virulent forms. As soon as it appears in a flock, not only the infected animals, but the whole flock with which they have been in contact, should at once be thoroughly well dipped, and as many of the "rubbings" disinfected with a solution of carbolic acid and water as it is possible to overtake. This operation is then to be repeated in the course of a week, and again in sixteen days, so as to effectually destroy the acari afterwards hatched. The frequent application of a properly prepared dip is thoroughly effective in destroying the living as well as the embryo parasite; and it is only through bad management or neglect that calamitous results need afterwards be feared.

Fuller information as to the most suitable "dips" to use is given under "DIPPING," page 140, to which the reader is referred. The following receipt, however, is recommended by Mr. James Archibald of Overshiels as being highly efficacious :—"To 40 gallons of water add 1 gallon of spirits of tar, 5 lbs. tobacco paper (infused), 5 lbs. soft soap, and 5 lbs. soda for 50 sheep." In mild cases one bath will be sufficient, and the most virulent forms completely cured by a second dose.

Scab is one of those contagious disorders which, on its appearance on a farm, requires to be reported to the local authorities, neglect to do so being punishable by law. Article 38 of the Animals Order of 1875 provides that "a local authority may from time to time, with a view of preventing the spreading of sheep-scab, make regulations for prohibiting a person from having in his possession or under his charge a sheep affected with sheep-scab without treating that sheep, or causing it to be treated, with some dressing or dipping, or other remedy for sheep-scab."

Foot-rot is to be met with on soils of every description; but sheep fed on dry, hard pastures, are less subject to the disease than those fed on soft and rank pastures. The finest and richest old pastures and lawns are particularly liable to it. Soft, marshy, and luxurious meadows are equally so; and it is also found in light, soft, or sandy districts. In the first of these it is perhaps most prevalent in a moist season, and in the latter in a dry one. On arable land it appears in young grass or seeds, if rich or luxuriant. In short, it exists to a greater or less extent in every situation which has a tendency to increase the growth of the hoofs without wearing them away, and more especially when they are kept soft by moisture. Soils of limestone formation, though not totally exempt from the disease, are more free from it than any other.

Some breeds of sheep are more susceptible to foot-rot than others, so much so, indeed, that some breeds—the Hampshire Down, for example—are popularly credited with a hereditary predisposition to the disease. No breed, however, is exempt from foot-rot; and if the Scotch black-faced breed suffers less from it than any other breed, it is perhaps largely owing to the fact that the blackfaces are accustomed to traverse a wider range, and that their native hills are, for the most part, dry, hard, and heathery.

T

There is a very general belief amongst shepherds that foot-rot is contagious. There are, however, several diseases of the foot of the sheep which are distinctly local and non-contagious—diseases which are due to injuries of different kinds inflicted during long journeys or by contact with irritating agents. Professor Brown, one of the best authorities on this point, says:—"Referring to the contagious character of foot-rot when sheep were kept in contact with diseased animals on moist and foul litter, the disease was communicated to them after several weeks' exposure. The animals suffering from the worst form of the disease were put with healthy sheep on perfectly dry ground on which a quantity of old building materials—brick rubbish and old mortar—were scattered. Instead of the disease extending, the diseased animals got well with extraordinary rapidity—a fact which is quite opposed to the idea that the disease possesses any high degree of the contagious property. It is not difficult to understand that when diseased sheep are put on moist pastures the infectious matter may be smeared on the long grass and be transferred to the skin between the digits of healthy sheep which pass over the same ground, and in such circumstances the conditions are most favourable for the development of the disease; but we have not met with that extremely infectious variety of the disorder which is communicated by the accidental contact of sheep in a market, or by merely passing along the road which has just been traversed by sheep affected by foot-rot. So far as we have seen, foot-and-mouth disease is the only affection of the foot of the sheep which can be transmitted to healthy animals so rapidly."

The diversity of opinion which exists amongst both practical men and veterinarians as to the contagious nature of foot-and-rot can only arise from confounding two diseases of the foot, somewhat similar in their early symptoms, but widely different in their effects. One of these affections is

due to the occurrence of some injury to the hoof-horn in the
first instance, so as to cause openings through which dirt
can obtain an entrance into the sensitive structure in the
interior of the hoof. This form of disease is most frequently
met with on sandy and chalky soils. An American writer
ascribes this form of the disease to the clogging up of the
biflex canal, thereby producing an active suppurating inflammation, which, if allowed to continue, soon becomes
chronic, extending its destructive ravages to all the tissues
of the foot. The biflex is described as a duct leading
from secretory gland of the foot to the base or divisions of
the hoof. Foot-rot of a more virulent kind than that which
is simply mechanical in its origin is found in most pastures
where the herbage is abundant and long. Instead of
beginning with an injury to the hoof, this form of the
disease arises from inflammation of the skin between the
digits or toes, and from this point the disease extends
downwards to the secreting membrane of the foot. It is
held by some that this form of foot-rot is of parasitic origin,
as seems highly probable ; but as yet all efforts to find the
foot-rot parasite have resulted in failure.

It has been suggested that the provisions of the Contagious
Diseases (Animals) Act should be extended to foot-rot in
sheep. This proposition was made seven years ago, and
the Privy Council was memorialised on the subject; but
there were practical difficulties in the way of the proposal.
In the first place, few farmers throughout the country were
favourable to impose restriction on the sheep trade on
account of foot-rot; and in the second place, veterinarians
were not only not agreed as to the contagious nature of the
disease, but even those who agreed thus far were frequently
unable to decide whether or not a disease in the foot of
a sheep was "contagious foot-root" or something else which
is not contagious. The *Veterinarian* and the *Veterinary
Journal* took different views on the subject—the former

contending that although the foot-rot may be communicated from a diseased to a healthy sheep by contact, it can only be transferred when the conditions are favourable, wet weather or wet lands being one of the conditions.

True foot-rot is distinguished in the early stage by an eruption on the surface of the skin between the claws; and the discharge which issues from the diseased surface will, if rubbed on the skin between the claws of a healthy sheep, produce the same disease, as may be proved by experiment. Therefore the popular notion as to the infectious character of foot-rot is not without foundation, although the disease cannot be transmitted without considerable difficulty. Foot-and-mouth disease, which rarely affects the mouth of the sheep, may be distinguished from foot-rot in the early stage, as it affects the skin of the heels immediately above the hoof-horn, instead of the part between the claws; but in the more advanced stages of the diseases, as when the whole foot has become affected, it is difficult even for a professional examiner to determine the precise nature of the malady. Some light may always, however, be thrown on the case by an inquiry into the history of the outbreak. Foot-and-mouth disease occurs suddenly; the animals are seen to be lame; they lose appetite for a time; and then, without any treatment, the majority of the sheep attacked gradually recover, chronic disease of the foot structure remaining in only a few. Foot-rot, on the other hand, does not appear to any extent all at once and without warning. A few animals show signs of lameness, which gradually increases as the disease in the foot advances, and there is no tendency to recovery so long as the sheep are kept under the conditions with which the disease is naturally associated—wet weather on a moist and gritty soil. It should also be noted that frequently in foot-and-mouth disease all the feet are affected, which in foot-rot is very rarely the case. Then in foot-and-mouth disease there are vesicles or blisters at the back of the foot, just where the horn joins the skin, or on the side of the foot in the same position. If the blisters have burst before being noticed, the white skin remains, and underneath it is red, raw surface of exposed vascular membrane—quite unlike the spongy fungoid structure which is seen under the ragged horn in foot-rot.

The treatment to be adopted will be very different in the two cases. Foot-rot requires the application of caustics to the fungoid granulations, and the removal of the animal to a dry situation, though in the worst forms of the disease complete cures have been effected without any medical applications by simply placing the sheep in a dry yard strewn with quicklime or thickly covered with old mortar. In foot-and-mouth disease nothing of the kind is required, as the feet, if kept free from dirt, will soon recover—that is, if the disease is detected at once; otherwise, if the suppuration of the hoof has been allowed to go on unnoticed, it will be necessary to remove the loose horn and apply caustic, as in ordinary foot-rot. But no disease or injury of the foot should be allowed to go on to this condition. The first sign of lameness should be accepted as an indication of disease, and at this early period there is usually little difficulty in discovering the cause and applying the proper remedies.

Nearly every shepherd has a nostrum of his own compounding for the treatment of foot-rot, all of them containing powerful caustics, such as crude carbolic acid, solution of nitrate of mercury, or the old butter of antimony diluted with tincture of benzoin. Professor Brown recommends a very effective caustic solution made by mixing together equal parts of solution of chloride of zinc and common hydrochloric acid. The diseased horn is first carefully cut away without bleeding; then the fungous growths are freely moistened with the caustic, after which a dressing of tar is sometimes supplied, though this is not necessary.

Paring and dressing require to be repeated at least once a week for a time. Where a large number of animals have to be dealt with, the method of driving them through a shallow trough in which a strong astringent solution is placed is sometimes adopted; but this is of no avail unless the feet are cleaned and pared. The solution, which should not be more than 1½ inches deep in the trough, may consist of 1 lb. of blue vitriol dissolved in 1½ gallons of water. Whatever dressing is used the sheep should not immediately thereafter be turned adrift on wet pasture or soft land, but kept standing for a few hours in a dry yard or shed, the floor of which is covered with quicklime or old mortar.

In addition to the treatment of the hoof itself it is always well to give internally the following :—

	Dram.
Common salt	1
Sulphate of iron	½
Nitrate of potash	½

Mix and give once a day.

Ticks and Keds.—The *Ixodes ricinus*, or sheep-tick, is of a whitish colour. They breed in the grass and attach themselves to the sheep in the months of April, May, and June, for the purpose of blood-sucking. Their effects, while not of a deadly nature, no doubt cause considerable discomfort and loss to the sheep.

When they first crawl on to the sheep they are comparatively small, but when distended with blood drawn from the sheep they are six times their original size. As soon as the parasite is filled almost to bursting, it quits its hold upon the sheep and drops to the ground, where it subsists possibly for a whole year upon the meal. This parasite is also supposed to give rise to the disease amongst sheep known as louping-ill. It has not been proved that such is actually the case, still there seem to be good reasons for justifying that belief. Ticks and louping-ill are synonymous to the same land; and while there may be ticks without louping-ill, there is never louping-ill without ticks. The *Ixodes ricinus* is only worth mentioning here on account of its surmised relation to this disease, and it may be hinted to shepherds and farmers in general not to fail to mark well its nature and habits, and to try if possible and discover how much truth there is in the tick theory.

The *Melophagus ovinus*, or sheep "ked," on the other hand, is a reddish parasite which breeds exclusively upon the sheep, not on the ground, as does its prototype the tick. The "ked" is found more or less in every flock, and if not destroyed by means of dipping, the "keds" become very numerous, and, by the irritation which they excite, cause a great loss of wool, from the sheep tearing the fleece with its teeth, or rubbing against the fences, not to mention the loss which accrues by the animal itself not thriving when subjected to continual torment. It is furnished with a sharp proboscis for blood-sucking, and when so engaged does so very vigorously, as if preparing for a long fast. Unlike the tick, however, after it has feasted it still retains its hold upon the sheep; but it is our opinion that a "ked," like the tick, does not feed more than once or twice in the year, even although it does remain upon the sheep. At any rate, that they can exist for a long period without fresh blood is proved by the fact of their being found alive in the shorn fleece twelve months after being removed from the sheep. They propagate by single births, the female laying a series of eggs during the season. At first these eggs are soft and of a yellow colour, but they soon harden, when they become dark brown. The eggs are held in the wool by the secretion from the skin of the sheep. They remain for about six weeks in this chrysalis state before becoming active, and in about as many more weeks the young are also reproductive. Careful dipping is the only effectual remedy for "keds."

Flies and Maggots.—In hot summer weather, and more especially during the months of July and August, sheep are subject to severe annoyance from the attacks of several kinds of flies. The loss and injury which these pests inflict upon the condition of sheep if definitely ascertained would reach a figure so far beyond ordinary belief, that even practical shepherds might reasonably doubt the truth of it; but let any one take notice of a flock of sheep on a hot summer's day, and observe the intense agony they are suffering from the torment of flies, and an approximate idea may be formed of the damage inflicted. Instead of quietly feeding or resting, the sheep are huddled together in a heap, pushing and crowding each other until they are almost suffocated. The heat exhaled from their own bodies, together with the rays of the sun, causes them to pant and gasp in a way suggestive of such torture, that it is plainly evident, however well the sheep may be fed, they cannot possibly under the circumstances be making good use of the food they eat, either in the early morning or in the cool of the evening; and when this daily torture is prolonged perhaps for weeks or months, it is a great deal more probable that the sheep will lose rather than gain in condition. So long, therefore, as the sheep are not improving, they are consuming food to no purpose, and every ounce thus eaten without yielding a proportionate increase is in consequence wasted. The loss sustained among cattle from the attacks of the gadfly has of late been brought prominently before the public both by new books on the subject and in agricultural newspapers, but little or nothing has as yet been said of the losses among sheep from pests of the same kind. Practical authorities have estimated the annual loss caused by the gadfly of the ox at several millions sterling, which has been the means of bringing out remedial suggestions from ento-mologists and others interested in the matter; and perhaps a similar calculation and discussion among those interested in sheep-farming may be the means of drawing scientific attention to the better protection of our flocks from the ravages of flies, which also so cruelly tease them in summer.

In estimating the damage which sheep suffer from flies, we have merely to reckon the injury sustained at the time of attack, as by proper management no other injury need follow.

In cattle, without taking into account the after effect and loss caused by the warble in the hide, it is said that the loss from attack and fright alone, in causing the animals to run and otherwise fret themselves, is not less than £1 per beast. How much then is the loss from the attack of flies on a single sheep? If we compare sheep with cattle grazing in the same field, it will appear to almost any observer that the sheep are suffering the most. At times the cattle may go off on a stampede and do themselves great injury, espe-cially if milch cows or fatting beasts; but they are seldom seen to be kept in the same agony all day long, as would a flock of sheep, by the myriads of flies which hover round them.

The smaller species of flies do comparatively little harm to cattle, but even the smallest midge seems to be a terror to sheep. Sheep, from being thinner-skinned and more easily frightened, have the most enemies, and their self-imposed crowding on warm days also affects them injuri-ously in a manner from which cattle are exempt. "Fly-time" lasts on an average about six weeks; and supposing the 32,000,000 of sheep in the country are fed for that length of time without a profit, and the grazing of each sheep costs 4d. per week, we get a total loss of £3,200,000—a sum certainly not overdrawn, yet sufficiently large to de-mand our best efforts to save. The species of flies most troublesome to sheep are the sheep-bot fly (*Œstrus cepha-lemyia ovis*), and the gadflies (*Tabanidæ*), of which there are many kinds closely resembling each other in their habits. "The *Œstrus ovis*," says Miss Omerod, "is rather larger than the common house-fly, and of an ashy colour spotted with black. The female either lays her eggs or deposits living maggots on the margins of the nostril of the sheep, from whence the maggots crawl up the nostrils by means of the mouth-hooks with which they are furnished, and attach themselves to the membranes of the cavities. Here they feed on the mucus; and it is stated they at times feed on the membrane itself, and also at times penetrate into the brain. Their presence causes great irritation, and when the attack is severe leads to gradual loss of strength and convulsions, ending in the death of the animal. In the common course of things the maggots remain in the head of the sheep for about eight to ten months before they are mature. They then leave the animal by going down the nostrils and fall to the ground, where they turn—either amongst roots or grass, or in any convenient place above or below the surface—to a black pupa, from which the fly comes out after a variable number of days, according to the climate." Sheep naturally protect themselves from the attack of the bot-fly by holding their nostrils down to the ground in places where dust or dry earth can be stirred in the action of breathing.

It appears that if the sheep can keep its nostrils perfectly dry the fly will not effect a lodgment, and on this account dust is absolutely essential as a protection. An American surgeon recommends as a preventive a cap made of lamb-skin tied around the sheep's head, which covers the face and has a fringe of strings hanging down over the nostrils, yet not interfering with the act of grazing. The strings or fringe is smeared with crude carbolic-acid ointment, made viscid like varnish with an addition of resin, and the short wool retains the odour of the ointment for a week or more. Should the fly effect a lodgment, a known remedy is to inject into the nostrils a teaspoonful of a mixture of equal parts of turpentine and linseed-oil, which either kills the grubs or causes them to be ejected by sneezing.

The "gadflies" (*Tabanidæ*) include some of the largest species of flies in this country, and are commonly known as flesh-flies, sheep maggot-flies, or blue-bottles. These flies are far more numerous and annoying to sheep than the bot-fly. The maggot-fly settles on the sheep about the root of the tail, or any other damp part of the fleece, where it deposits its eggs. It has been ascertained that one female will lay as many as 20,000 eggs, which hatch so quickly that in twenty-four hours the larvæ will devour so much food and grow so fast as to increase their weight two hundred-fold. The maggots soon become active, and spreading from their quarters, attack the skin, which they irritate and cause to secrete a serous fluid. In time the skin is pierced, and if the animal be not attended to, it will soon be found a lifeless mass covered with these loathsome vermin. A fly-blown sheep can readily be distinguished. It will stand perfectly still with the head near the ground, ears slightly drooping, when all of a sudden it will start to run, giving short quick jerks with the hind-legs, something like a horse that is string-halted, and a peculiar sideway motion to the spine.

This is the first indication of fly-blow the shepherd has. As the maggots increase in size and number the poor sheep will lie down, jump up with a start, utter a mournful bleat, as if it recognised that unless relieved soon its days would be few.

The best remedy for a fly-blown sheep is to take of kerosene and water equal parts, and bathe the parts infected with maggots. The following receipts are also useful aplications:—(1.) Turpentine, 1 part; olive oil, 3 parts.

(2.) Mercury, ½ oz. ; turpentine, ½ pint ; tincture of myrrh, ½ oz. ; friars balsam, ½ oz. ; and put two table-spoonfuls into a quart bottle of water for application.   This will usually kill them instantly.   Watch the sheep, and if any maggots have escaped the first time, give another application in say from four to eight hours.   If this second application does not remove all, repeat every four or eight hours until no more maggots are to be found.

Then apply any healing ointment to the part where the maggots were.   If you have no ointment, apply good clean lard two or three times until the sores are entirely healed.   To prevent the fly troubling sheep on the head, make a plaster of pitch and bee's wax, or thick tar and sulphur, and smeer the affected part.   As a general protection against all flies and ticks from which sheep are liable to suffer, there is no better agent than the usual process of dipping.   Of course dipping merely once or twice a year has little effect in preventing the attack of flies ; it requires to be done often, and by using a weak solution it will pay to dip the sheep every fortnight during the fly season.   It is a perfect pleasure to see how differently the sheep feed unmolested with flies after they have been bathed.   But although dipping is very useful in its way, we want some better fly-disguster than any present sheep dip contains.   There is plenty of room for improvement on dipping compounds, and manufacturers must not for a moment suppose they have already produced the desired article.

Awalding.—Though not a disease, "awalding," or lying awkward, is the most common and dangerous of sheep accidents.   It occurs most frequently when the weather grows warm and sultry after a shower—from the beginning of May until the sheep are shorn, after which the danger is over ; this raising an itching in their backs—due to vermin principally—they lie down and roll, and when it happens to be a level place, or in a furrow, owing to the bulk of their fleeces, they cannot get up again; and, with the stomach full of food, may die almost immediately from apoplexy. The shepherd cannot help sheep falling awkward, but it is a disgrace to him to let one die in this way. Some members of a flock are habitual offenders in this way, and cause constant watching and lifting. It is sometimes necessary to confine these individuals where they can be more conveniently seen, than when left on distant moors. This plan is, at any rate, preferable to the practice which inhuman shepherds adopt of paring the sheep's feet to the quick.

CHAPTER XXIX.

CONCLUSION.

Dentition.—There are two sets of teeth in the sheep: a first deciduous or milk dentition, and a second permanent or adult dentition. Each set when complete consists of incisor, canine, and molar teeth. The permanent teeth which replace the deciduous molars are called premolars. Behind the milk molars, three grinding teeth on each side of each pair are developed, and these come into place without replacing any other teeth from below. These are called the true molars.

Professor Huxley has employed a very simple and convenient method of showing the number and disposition of the teeth. The letters *di, dc, dm,* are made to represent, respectively, the deciduous or milk set of incisors, canines, and molars. The same letters without the prefix *d* are used for the permanent incisors, canines, and molars, with the exception that the premolars have the symbol *pm.* Then, by placing after each of these symbols figures arranged so as to show the number of teeth of each kind symbolised, on each side of each jaw, we have the *dental formula* of the animal. In the sheep the permanent dentition is represented thus :—

	Incisors.	Canines.	Pre-Molars.	Molars.	Total.
Upper jaw	0—0	0—0	3—3	3—3	= 32
Lower jaw	3—3	1—1	3—3	3—3	

In sheep the milk or lamb teeth are easily distinguished

from the permanent or broad teeth by their smaller size, and by the thickness of the jaw-bone around their fangs, where the permanent teeth are still enclosed. As the lamb approaches a year old, the broad exposed part of the tooth becomes worn away, and narrow fangs projecting above the gums stand apart from each other, leaving wide intervals. This is even more marked after the first pair of permanent teeth have come up, over-lapping each other at their edges, and from this time onward the number of small milk teeth and of broad permanent teeth can usually be made out with ease. Another distinguishing feature is the yellow or dark coloration of the fangs of the milk teeth, while the exposed portions of the permanent teeth are white, clear, and pearly.

The age of a sheep can be determined by the incisor or front teeth. At a month old a lamb will have eight temporary incisors or milk teeth. At from twelve to fifteen months old the centre pair of milk incisors will be replaced by two larger permanent ones. At eighteen months to two years the second pair of permanent incisors will have appeared; the third pair at two years and six months, and the fourth and last pair at three years, or shortly after. In the case of corn and turnip fed sheep the dentition is some months earlier.

Hill sheep retain their teeth complete and unbroken until about six or seven years of age. After that period the teeth are liable to decay, although many of the teeth keep good in their mouths for some years longer. It all depends on how the sheep are fed for destroying the teeth at an early age. Turnip fed sheep lose their teeth very quickly, while hill flocks that feed on grass only may keep them good in their mouths for an indefinite number of years.

INDEX.

———◆———

U

PRINTED BY BALLANTYNE, HANSON AND CO.
EDINBURGH AND LONDON.

www.ingramcontent.com/pod-product-compliance
Lightning Source LLC
Chambersburg PA
CBHW081715220526
45468CB00008B/1857